Photodiode Amplifiers

Photodiode Amplifiers

Op Amp Solutions

Jerald G. Graeme
Principal Engineer
Gain Technology Corporation

McGraw-Hill

New York San Francisco Washington, D.C. Auckland Bogotá
Caracas Lisbon London Madrid Mexico City Milan
Montreal New Delhi San Juan Singapore
Sydney Tokyo Toronto

Library of Congress Cataloging-in-Publication Data

Graeme, Jerald G.
  Photodiode amplifiers : op amp solutions / Jerald G. Graeme.
    p.   cm.
  Includes bibliographical references and index.
  ISBN 0-07-024247-X (hc)
  1. Operational amplifiers—Design and construction. 2. Diodes, Semiconductor.   3. Photoelectric cells.   I. Title.
  TK7871.58.06G733 1995
  621.39′5—dc20   95-42800
                                                                  CIP

McGraw-Hill
A Division of The **McGraw·Hill** Companies

Copyright © 1996 by The McGraw-Hill Companies, Inc. All rights reserved. Printed in the United States of America. Except as permitted under the United States Copyright Act of 1976, no part of this publication may be reproduced or distributed in any form or by any means, or stored in a data base or retrieval system, without the prior written permission of the publisher.

1 2 3 4 5 6 7 8 9 0   QBP/QBP   9 0 9 8 7 6 5

ISBN 0-07-024247-X

The sponsoring editor for this book was Steve Chapman, the editing supervisor was Fred Bernardi, and the production supervisor was Pamela Pelton. Illustrated by Lola E. Graeme. It was set in Century Schoolbook by ETP/Harrison.

Printed and bound by Quebecor-Book Press

McGraw-Hill books are available at special quantity discounts to use as premiums and sales promotions, or for use in corporate training programs. For more information, please write to the Director of Special Sales, McGraw-Hill, 11 West 19th Street, New York, NY 10011. Or contact your local bookstore.

Information contained in this work has been obtained by The McGraw-Hill Companies, Inc. ("McGraw-Hill") from sources believed to be reliable. However, neither McGraw-Hill nor its authors guarantees the accuracy or completeness of any information published herein and neither McGraw-Hill nor its authors shall be responsible for any errors, omissions, or damages arising out of use of this information. This work is published with the understanding that McGraw-Hill and its authors are supplying information but are not attempting to render engineering or other professional services. If such services are required, the assistance of an appropriate professional should be sought.

Contents

Preface   ix

Chapter 1. Photodiodes   1

1.1   The Photo Effect   1
1.2   The Photodiode Model   4
1.3   Photodiode Variations   7
    1.3.1   PIN Photodiodes   7
    1.3.2   Avalanche Photodiodes   9
1.4   Position-Sensing Photodiodes   10
    1.4.1   The Basic Lateral Photodiode   11
    1.4.2   The Lateral Photodiode Model   13
    1.4.3   The Duo-Lateral Photodiode   15
    1.4.4   The Tetra-Lateral Photodiode   17
    References   19

Chapter 2. The Basic Amplifier   21

2.1   Linearity   21
2.2   Offset   24
    2.2.1   Reducing Offset with a Compensation Resistor   24
    2.2.2   Reducing Offset with a Feedback Tee   24
2.3   Bandwidth   28
    References   30

Chapter 3. Bandwidth and Stability   31

3.1   The Inherent Response Limits   32
    3.1.1   The Response Limit of Parasitic Capacitance   32
    3.1.2   The Response Limit of Op Amp Bandwidth   33
3.2   The Phase Compensation Requirement   36
    3.2.1   The L-C Tank Equivalent and Phase Compensation   36
    3.2.2   Feedback Analysis of the Basic Circuit   38
3.3   Phase Compensation   42

vi Contents

        3.3.1 The Basic Compensation   42
        3.3.2 Phase Margin Analysis   43
        3.3.3 Phase Margin Analysis of the Typical Case   45
        3.3.4 Phase Margin Analysis of the Universal Case   47
        3.3.5 Selecting the Phase Compensation Capacitor   49
   3.4 The Bandwidth Advantage of the Current-to-Voltage Converter   50
   3.5 Phase Compensation Alternatives   53
        3.5.1 The Photodiode Amplifier's Second-Order Response   53
        3.5.2 Translating the Second-Order Results to the Photodiode Amplifier   56
        3.5.3 Selecting the Phase Compensation for Limited Peaking   57
        3.5.4 Realizing the Precise Phase Compensation   59
   References   61

Chapter 4. Wideband Photodiode Amplifiers   63

   4.1 Biasing the Photodiode   63
        4.1.1 The Bias Effect   64
        4.1.2 Photodiode Biasing with the Current-to-Voltage Converter   65
   4.2 Bias Improvements   67
        4.2.1 Filtering the Bias Voltage   67
        4.2.2 Common-Mode Rejection of the Bias-Induced Errors   69
   4.3 Bootstrapping the Photodiode   71
        4.3.1 The Basic Bootstrap Photodiode Amplifier   71
        4.3.2 Bandwidth Analysis of the Bootstrap Amplifier   74
        4.3.3 Phase Compensating the Bootstrap Amplifier   74
   4.4 Combining Bootstrapping with the Current-to-Voltage Converter   77
        4.4.1 The Basic Combination and Its Requirements   77
        4.4.2 A Practical Buffer Solution   80
        4.4.3 Bandwidth Analysis of the Combination   83
   References   85

Chapter 5. Noise   87

   5.1 General Noise Effects   87
        5.1.1 Noise Densities and Noise Gains   88
        5.1.2 Noise Gain Peaking   90
        5.1.3 Total rms Output Noise   92
   5.2 The Effect of Op Amp Input Noise Voltage   94
        5.2.1 An Intuitive Derivation of the e_{noe} Noise Component   94
        5.2.2 Simplifying the E_{noe} Analysis   95
        5.2.3 The E_{noe} Components   97
   5.3 Combining the Noise Effects   99
        5.3.1 The Noise Analysis Summary   99
        5.3.2 Identifying the Predominant Noise Effect   102
   References   106

Chapter 6. Noise Reduction   107

   6.1 Noise Reduction with Feedback Capacitance C_f 108
        6.1.1 The Noise Gain Reduction of C_f 108
        6.1.2 Noise Analysis for the C_f Case   110

6.2	Noise Bandwidth versus Signal Bandwidth	112
6.3	Noise Reduction with a Composite Amplifier	112
	6.3.1 The Noise Bandwidth Reduction of the Composite Amplifier	112
	6.3.2 The Noise Gain Reduction of the Composite Amplifier	115
	6.3.3 Optimizing the Composite Noise-versus-Bandwidth Compromise	117
	6.3.4 Noise Analysis for the Composite Case	118
	6.3.5 A Comparison with the Active Filter Alternative	120
6.4	Noise Reduction with Decoupling Phase Compensation	121
	6.4.1 The Noise Gain Reduction of the Decoupling	122
	6.4.2 Optimizing the Decoupling Noise-versus-Bandwidth Compromise	123
	6.4.3 Noise Analysis for the Decoupling Case	126
	References	128

Chapter 7. High-Gain Photodiode Amplifiers — 129

7.1	Using a Feedback Tee Network	130
	7.1.1 The Gain and Noise Produced by a Feedback Tee	130
	7.1.2 Optimizing the Tee's Noise Voltage Response	134
	7.1.3 Optimizing the Tee's Resistance Noise Response	135
	7.1.4 Noise Analysis for the Tee Case	138
7.2	Adding a Voltage Amplifier	140
	7.2.1 Optimizing the Two-Amplifier Bandwidth-versus-Noise Compromise	141
	7.2.2 Designing for the Bandwidth-versus-Noise Optimum	144
	7.2.3 Noise Analysis for the Two-Amplifier Case	146
7.3	Adding Voltage Gain	150
	7.3.1 Voltage Gain Alternatives	150
	7.3.2 Optimizing the Single-Amplifier Bandwidth-versus-Noise Compromise	152
	7.3.3 Noise Analysis for the Single-Amplifier Case	155
7.4	Adding Current Gain	157
	7.4.1 Current Gain Alternatives	157
	7.4.2 Optimizing the Current-Gain Bandwidth-versus-Noise Compromise	161
	7.4.3 Noise Analysis for the Current Gain Case	163
	References	164

Chapter 8. Reducing Power-Supply Noise Coupling — 165

8.1	The Power-Supply Bypass Requirement	166
	8.1.1 The Noise-Coupling Mechanism	167
	8.1.2 The Noise-Coupling Frequency Response	168
	8.1.3 Power-Supply Coupling and Frequency Stability	170
	8.1.4 The Oscillation Condition	172
8.2	Selecting the Primary Bypass Capacitor	173
	8.2.1 Bypass Resonances	174
	8.2.2 An Intuitive Evaluation of Bypass Resonance	176
	8.2.3 The Bypass Selection	177
8.3	Selecting a Secondary Bypass Capacitor	179
	8.3.1 Bypass Capacitor Self-Resonance	179

viii Contents

		8.3.2 Dual Bypass Capacitors	182
		8.3.3 Dual Bypass Selection	183
	8.4	Bypass Alternatives	185
		8.4.1 Detuning the Dual Bypass Resonance	186
		8.4.2 Selecting the Detuning Resistance	188
	8.5	Power-Supply Decoupling	191
		8.5.1 Decoupling Alternatives	191
		8.5.2 Selecting the Decoupling Components	195
		References	196

Chapter 9. Reducing External Noise Effects 197

	9.1	Reducing Electrostatic Coupling	198
		9.1.1 Electrostatic Shielding	198
		9.1.2 The Differential-Input Current-to-Voltage Converter	199
		9.1.3 Other Noise Effects of the Differential-Input Connection	202
		9.1.4 An Alternate Differential-Input Photodiode Amplifier	207
		9.1.5 Other Noise Effects of the Differential-Input Alternative	209
	9.2	Reducing Magnetic and RFI Coupling	210
		9.2.1 Magnetic Shielding	210
		9.2.2 Circuit Reductions of Magnetic Coupling	212
	9.3	Reducing Multiple Coupled-Noise Effects	216
	9.4	Minimizing Magnetic Field Generation	216
		References	219

Chapter 10. Position-Sensing Photodiode Amplifiers 221

	10.1	Direct-Displacement Monitor Amplifiers	222
		10.1.1 Single-Axis Monitoring with a Differential-Photodiode Amplifier	222
		10.1.2 Noise Reduction with the Differential-Photodiode Amplifier	224
		10.1.3 Connecting the Photodiodes Differentially	225
		10.1.4 An Alternate Differential-Photodiode Connection	227
		10.1.5 Two-Axis Monitoring	229
	10.2	Normalized Monitor Amplifiers	231
		10.2.1 Normalizing the Differential-Photodiode Amplifier	232
		10.2.2 Easing the Normalization	234
		10.2.3 Simplifying the Normalized Differential-Photodiode Amplifier	237
	10.3	A Digital Alternative to Normalization	239
		10.3.1 Processing the Array Signals	239
		10.3.2 Defining the Logic Output Levels	241
		References	243

Glossary 245

Index 249

Preface

Photodiodes transform a basic physical indicator, light, into the electrical form commonly used to monitor physical conditions. There, semiconductor junctions convert the photon energy of light into an electrical signal by releasing and accelerating current-conducting carriers within the semiconductor. Specializing the junction design for the photodiode role improves its spectral response and efficiency with PIN and avalanche alternatives. Also, multiple-element and lateral photodiodes provide position-sensing through the relative magnitudes of multiple output currents. By itself, the photodiode can produce a voltage output as required for most electronic instrumentation. However, that operating mode produces a highly nonlinear response and a very restricted bandwidth. Instead, accepting the diode output as a current and performing a current-to-voltage conversion dramatically improves performance. This isolates the photodiode from the signal voltage and makes the op amp current-to-voltage converter the basic photodiode amplifier.

While simple in structure, this circuit still displays surprising multidimensional constraints in photodiode applications. There, the small output levels produced by the photodiode encourage large-area diodes and high-value conversion resistances. However, this combination compromises offset, bandwidth, stability, and noise. These compromises require careful design choices and suggest variations on the basic photodiode amplifier. In the chapters that follow, design equations guide the circuit's component selection, and circuit derivatives optimize the various performance characteristics. In the first circuit variation, replacing the current-to-voltage conversion resistance with a tee network reduces offset, and a later analysis guides the tee design to avoid the customary noise increase of this replacement.

Next, the high resistance of the conversion resistor and the capacitance of the photodiode combine to produce three separate frequency-response limits. Each potentially sets the bandwidth of the current-to-voltage converter when operated as a photodiode amplifier. These limits result from parasitic capacitance, the limited bandwidth of the op amp, and phase compensation requirements. In the first case, parasitic capacitance bypass of the large conversion resistance rolls off the circuit's response. In other cases, the op amp bandwidth and phase compensation compete to control bandwidth. As a photodiode amplifier, the current-to-voltage converter often produces a two-pole response, resulting from an input circuit that appears like a parallel L-C circuit. This characteristic jeopardizes stability and requires the phase compensation that introduces the third response limit. Feedback analysis develops design equations for this compensation in a compromise that equates the bandwidth limits imposed by the op amp and this compensation.

Reducing the effects of the photodiode capacitance greatly improves bandwidth. Any signal voltage developed across the diode reacts with this capacitance, shunting the diode's output current at higher frequencies. Three circuit methods greatly ease this restriction through signal isolation, photodiode bias, and photodiode bootstrap. Signal isolation removes the signal voltage from the photodiode through the current-to-voltage converter of the basic photodiode amplifier. For even greater bandwidth, biasing or bootstrapping the photodiode further reduces the capacitance effect. Reverse biasing the diode reduces its capacitance but requires additional circuit complexity to minimize the accompanying degradation of offset and noise performance. Bootstrap also isolates the photodiode from signal swing, and combining this with the current-to-voltage converter provides a double degree of isolation.

The photodiode amplifier's resistance, capacitance combination also degrades noise performance, and for higher-gain applications, this noise often controls the amplifier's overall accuracy. As a photodiode amplifier, the current-to-voltage converter typically exhibits a complex noise behavior in which the op amp's input noise voltage receives an unexpected high-frequency gain. Here, breaking the noise analysis into a series of frequency regions restores simplicity. This regional analysis also permits comparison of relative noise effects to identify circuit changes that optimize performance. Frequently, this analysis reveals that the unexpected high-frequency gain or noise gain peaking dominates noise performance. Then, the circuit amplifies the noise voltage of the op amp's input with a noise bandwidth that exceeds the signal bandwidth. Modifications to the basic photodiode amplifier provide filtering to reduce or remove this noise disadvantage. In each case, noise analysis equations guide the circuit's component selection to optimize the circuit's bandwidth-versus-noise compromise.

Bandwidth and noise also compete when maximizing the photodiode amplifier's gain. For the basic photodiode amplifier, just making the feedback resistance large produces high gain. However, the very high resistances often required reduce bandwidth and increase noise. For those cases, alternative methods optimize this compromise. Increased bandwidth results from supplemental gain provided by several circuit alternatives. The most straightforward implementation simply adds a voltage amplifier following the conventional current-to-voltage converter. A second reduces the circuit complexity by making one op amp serve both the current-to-voltage and voltage gain functions. Current output along with increased bandwidth result from replacing the added voltage gain with current gain. In each case, design equations guide component selection for the optimum bandwidth-versus-noise condition.

Noise on the power-supply lines also couples into the signal path of the photodiode amplifier. The finite power-supply rejection ratio of the op amp code couples a portion of this noise to the op amp inputs, and the circuit amplifies this added signal along with the op amp's normal input noise. There, the characteristic noise gain peaking of the photodiode amplifier makes supply noise reduction especially important. In addition, this gain peaking makes the photodiode amplifier more vulnerable to oscillation through a parasitic feedback loop formed through the power-supply lines. Fortunately, capacitive bypass of the power-supply lines greatly attenuates this noise coupling and ensures frequency stability as well. However, to be successful, the bypass selection requires close attention to the frequency-dependent impedances of both the supply lines and the bypass capacitors. Together, they produce multiple resonances in the net supply-line impedance that increase the gain of the parasitic feedback loop. Once again, design equations guide component selection for this bypass function.

Diminishing returns eventually limit the noise reduction achieved through measures focused upon the photodiode amplifier itself. External noise sources impose a background noise floor that requires attention to the amplifier's environment rather than the amplifier. This background noise typically results from the parasitic noise coupling of external electrostatic and magnetic sources. Here again, the noise gain peaking of the photodiode amplifier accentuates the effects of these noise sources. Limiting these effects requires attention to amplifier location, shielding, the circuit structure, and the physical arrangement of circuit components. There, differential amplifier structures and loop-minimizing component arrangements reduce the effects of these external noise sources. Converting the photodiode amplifier to a differential-input form activates the common-mode rejection of the op amp for rejection of both electrostatic and magnetic coupling effects.

In a final set of circuit alternatives, photodiodes provide position-sensing information through the diodes' photo responses. However, several variables potentially degrade the sensing accuracy by introducing offset and gain errors in the light-to-voltage conversion process. Background light introduces a first variable but only adds an offset to the diode output. Differential measurement with two photodiodes removes this offset through the subtraction of two output signals. Variations in the light source intensity and in the photodiode's responsivity introduce additional variables, and these affect the gain of the position measurement. Simple differential monitoring does not remove these effects. However, signal normalization resolves this by dividing two diodes' difference signal by their mean signal. An analog divider extracts this normalized signal with the greatest accuracy, but analog multiplier replacements for the divider simplify the circuitry. Alternately, linear photodiode arrays remove the need for normalization by providing a digital indication of position.

I wish to thank the many photodiode users with whom I have discussed applications requirements over the years. Their shared knowledge and catalytic inquiries prompted the investigations that led to this book. My thanks to *EDN* and *Electronic Design* magazines for publishing the articles that initiated many of these discussions. Finally, I wish to thank my wife, Lola, for her accurate and attractive rendering of the illustrations and for the rewarding feeling of mutual involvement in preparing this book.

Jerald Graeme

Photodiode Amplifiers

Chapter 1

Photodiodes

Semiconductor junctions convert the photon energy of light into an electrical signal by releasing and accelerating current-conducting carriers within the semiconductor. All semiconductor junctions display this response, and this forms the basis for the photodiode. Thus, a photodiode behaves like an ordinary signal diode but with the addition of an internally generated current derived from illumination. Specializing the diode design for the photodiode role improves its spectral response and efficiency. PIN photodiodes expand the response range through an added intrinsic layer formed between the normal P and N regions. Avalanche photodiodes increase efficiency through an internal current gain developed by avalanche multiplication. In every case, the magnitude of the photo-generated current corresponds directly to the light intensity, as required for most photodiode applications. Other applications monitor the relative magnitudes of multiple output currents to detect the position rather than intensity of a light beam. In the simplest case, multiple-element photodiodes develop the latter currents, but lateral photodiodes improve position-sensing resolution through multiple outputs taken from a single diode. There, resistive current division separates the lateral photodiode's current into multiple outputs.

1.1 The Photo Effect

Light entering a semiconductor material produces an electrical current by releasing hole-electron pairs as illustrated in Fig. 1.1. There, photons transfer energy to the atoms of the radiated material, moving these hole and electron carriers to their conduction states. Once there, the individual carriers may or may not contribute to current flow. Carriers released within the depletion region of a semiconductor

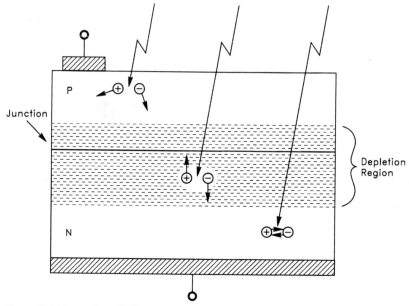

Figure 1.1 In a photodiode, holes and electrons released by photons within the depletion region accelerate toward the diode terminals, but those released outside this region travel by diffusion or simply recombine.

junction produce the majority of this current due to the electric field of this region. That region contains ionized or depleted atoms that support a voltage differential across the junction. The associated electric field accelerates the carriers toward the diode's top and bottom terminals, adding conduction energy to the carriers and reducing the probability of recombination. Applying reverse bias to the junction expands the depletion region to encompass more of the diode's material within the accelerating field. However, in the absence of external bias, a depletion region still develops at any semiconductor junction due to the junction's built-in voltage. There, just the diffusion of thermally generated carriers across the junction produces a region of exposed charge. At equilibrium, that diffusion produces a charge balance across the junction that supports an electric field. Carriers released away from the depletion region diffuse through the material until they reach that region or recombine. Those reaching the region receive the electric field's acceleration to be swept to the diode terminals as part of the conduction current. Typically, only those carriers generated within one diffusion length of the depletion region contribute to current flow.

The construction of the photodiode and the wavelength of the light strongly influence the efficiency of the light-to-current conversion. There, semiconductor doping levels and the junction depth become key parameters. Lightly doped materials expand the depletion region by reducing the number of doped atoms per unit volume. This doping, or selective impurity addition, produces atoms with potential hole or electron carriers residing near the conduction band energy level of the semiconductor material. At their higher energy levels, these atoms more easily ionize, and when ionized, they support the field of the depletion region. Then, a lighter doping level increases the depletion volume occupied by the exposed charge needed to support the field. However, the low resistivity required for electrical contact to the material generally limits this doping option.

The depth and extent of the junction determines the location of the depletion region and the light wavelengths that produce an efficient response. Photons generate carriers at a range of depths with a given range proportional to the photon wavelength. Thus, shallow junctions efficiently convert short wavelengths with a depletion region that easily encompasses the majority of the photo generation. However, longer wavelengths require increasingly wider depletion regions for efficient conversion, as addressed by the PIN photodiodes discussed later. For a given photodiode and a given wavelength, a photodiode responsivity expresses the resulting efficiency through[1]

$$i_p = r_\phi \phi_e$$

Here, r_ϕ is the diode's flux responsivity and ϕ_e is the radiant flux energy in watts.

The ac response of i_p displays a dual time constant due to the two carrier travel mechanisms that account for photodiode current.[2] As described above, carriers generated both within and outside the depletion region contribute to this current. Those generated within the region travel under the accelerating influence of the region's electric field and proceed rapidly to the diode's terminals. This produces a fast or drift component of i_p, i_{dr}, as controlled by the drift time of the depletion region. Those carriers generated outside the region initially travel with little influence imposed as to direction or speed and proceed slowly. When the latter carriers reach the depletion region, they too proceed rapidly, but the interim diffusion time produces a slow component of i_p, i_{di}. The combination of the two components produces a time domain current of $i_p(t) = \alpha_{dr} i_{dr}(t) + \alpha_{di} i_{di}(t)$, where α_{dr} and α_{di} represent the fractions of i_p supplied by the drift and diffusion components. This produces a step response

$$i_p(t) = i_p(\infty) \left(1 - \alpha_{dr} e^{-t/\tau_{dr}} - \alpha_{di} e^{-t/\tau_{di}}\right)$$

where the τ_{dr} and τ_{di} time constants characterize the drift and diffusion responses.

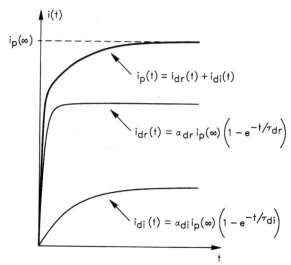

Figure 1.2 The drift and diffusion components of photodiode current exhibit greatly different time constants to produce a net current displaying a two-time-constant response.

Figure 1.2 illustrates these two responses and the composite $i_p(t)$ result. There, the shapes of the i_{dr} and i_{di} curves reflect their greatly different τ_{dr} and τ_{di} time constants. Adding the two curves produces the i_p curve with a characteristic two-time-constant shape. Initially, i_p rises rapidly due to the effect of i_{dr}, after which i_p assumes the long settling time produced by i_{di}. This settling time defines the fundamental ac response limit for the photodiode. However, most photodiode applications introduce more significant limits through the photodiode capacitance and a monitor amplifier bandwidth limit. Chapter 3 examines these other effects.

1.2 The Photodiode Model

Modeling the characteristics of the photodiode with discrete circuit components permits analysis of application circuits. Figure 1.3 shows the resulting model with an ideal diode, a current source, and accompanying parasitic elements. Current source i_p represents the photodiode signal, and the diode replicates voltage conditions for the forward-biased state. Resistance R_D represents the diode's dark resistance, which is the resistance of the zero-biased diode junction. For most applications, this high resistance produces little effect and can be ignored. Similarly, resistance R_S models the series resistance of the semiconductor material, and this low resistance can generally be ignored. However, the remaining parasitic element C_D produces profound performance effects for most photodiode applications. Later chapters describe the stability, bandwidth, and noise compromises produced by this capacitance.

C_D represents the stored charge effect of the photodiode junction and varies with the diode's area and voltage. Larger diode areas encompass a greater junction volume with increased stored charge and a corresponding larger value for C_D. Reverse biasing the diode increases the depletion width of the junction to effectively increase the space between the capacitor's plates and reduce C_D. The latter effect is described by

$$C_D = \frac{C_{D0}}{\sqrt{1 + V_R/\varphi_B}}$$

Here, C_{D0} is the photodiode capacitance at zero bias and φ_B is the built-in voltage of the diode junction. Applying a reverse bias voltage V_R reduces C_D from its C_{D0} value through a comparative relationship with φ_B. Chapter 4 quantifies the benefit of the reduced capacitance with practical circuit applications.

A simplified version of this model helps explain the characteristic curves measured with a photodiode. For a signal diode, sweeping the voltage impressed across the device and measuring the terminal current defines its characteristic curve. A photodiode simply introduces another variable through internally generated current. However, evaluation of the photodiode performance requires making a distinction between the terminal current measured and the photo-generated current. Figure 1.4 illustrates this difference with the diode and current source of Fig. 1.3. The simplified model here neglects the resistances of the earlier case since they generally produce negligible effects. Similarly, this model neglects the earlier capacitance for the low frequencies common to characteristic curve measurement. With this model, an impressed test voltage e_p produces a terminal current $i_T = i_d - i_p$

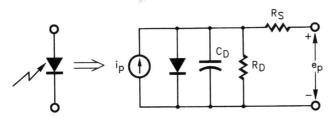

Figure 1.3 The circuit model of a photodiode consists of a signal current, an ideal diode, a junction capacitance, and parasitic series and shunt resistances.

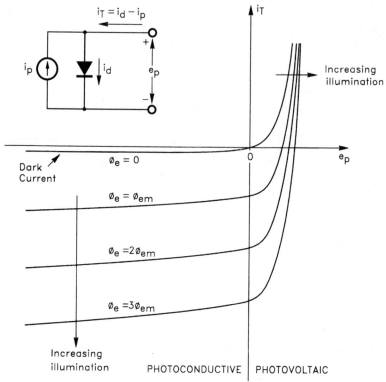

Figure 1.4 Sweeping the voltage and measuring the current with a photodiode reproduces the basic diode curve with offsets produced by the photodiode current.

and, here, a measurement-generated diode current i_d differentiates the measured i_T current from the photo-generated current i_p.

Depending upon the polarity of the impressed e_p, this difference may or may not influence the intuitive evaluation of the resulting characteristic curves. Figure 1.4 shows these curves for zero illumination flux, $\phi_e = 0$, and for several multiples of a measurement flux density ϕ_{em}. The result simply replicates the familiar current-voltage characteristic of a diode with varying levels of offset. However, interpreting the photo gain from these curves requires switching from vertical to horizontal comparisons as the test voltage e_p passes through zero. For $e_p \leq 0$, the diode of the model remains reverse biased, and $i_d = 0$ for $i_T = -i_p$. Then, the measured current directly reflects that generated by the diode's photo response. In this region of the curves, zero illumination, or $\phi_e = 0$, leaves i_p at its small leakage or dark current level. Then, increasing values of ϕ_e raise i_p linearly and step the $i_T = -i_p$ curves vertically downward in the figure. The vertical spacing between the curves reflects the photo gain and the photodiode responsivity r_ϕ of the earlier $i_p = r_\phi \phi_e$. Also in this region, the curves display slopes reflecting the increased photodiode responsivity as the e_p voltage moves more negative. As mentioned before, this reverse bias increases the width of the diode's depletion region to produce this increased responsivity. This $e_p < 0$ region represents the photoconductive or current-output region of the photodiode response.

For $e_p > 0$, the diode becomes forward biased and influences the current measured. Then, $i_d \neq 0$, and the forward bias increases the measured $i_T = i_d - i_p$. In this region of the curves, points along their paths no longer reflect just the photo response. However, the spacings between the curves still indicate the photo gain because i_p alone produces these. The horizontal spacing between the curves reflects the photo gain for the photovoltaic or voltage-output operating mode of the photodiode. In the actual photovoltaic mode, no e_p measurement signal drives the circuit. Instead, i_p supplies the total diode current, $i_d = i_p$ in Fig. 1.4, and produces the voltage e_p as an output signal. Then, the flow of i_p through the diode produces $e_p = V_t \ln(i_p/I_D)$, where $V_t = KT/q$ is the thermal voltage of the junction, and I_D is the dark current or reverse saturation current of the diode. From before, i_p varies linearly with illumination intensity ϕ_e as expressed by $i_p = r_\phi \phi_e$. This makes

$$e_p = V_t \ln \frac{r_\phi \phi_e}{I_D}$$

and e_p is a logarithmic rather than linear function of light intensity.

1.3 Photodiode Variations

Two variations of the basic photodiode improve the diode's response. PIN photodiodes increase the spectral bandwidth or range of light frequencies that produce an efficient photo response. Avalanche photodiodes increase the magnitude of the output current and the response speed by permitting diode bias at the verge of breakdown. However, control of this bias presents significant practical difficulties.

1.3.1 PIN photodiodes

Adding an intrinsic layer to the photodiode structure increases the spectral range of response by expanding the depletion region. As described in Sec. 1.1, photons release carriers in a semiconductor over a range of depths proportional to the photon wavelengths, and only those carriers released within or near the depletion region contribute

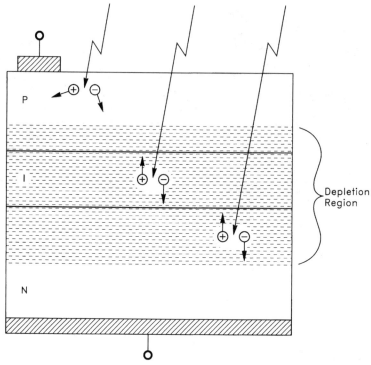

Figure 1.5 PIN photodiodes include an intrinsic interlayer that expands the depletion region to encompass carriers released by a broader range of photon wavelengths.

to the diode current. This makes the basic photodiode somewhat wavelength selective, and that feature benefits some applications. Other applications require broader spectral responses as provided by the PIN photodiode structure of Fig. 1.5. As the name suggests, this photodiode includes an intrinsic region between the P and N regions. While never truly intrinsic,[3] the added region contains relatively few doped and depletable atoms.

Then, to achieve charge balance, the depletion region must extend through the intrinsic region to the adjoining and heavier-doped P and N regions. As illustrated, this extends the span of the depletion region to encompass carriers released by a broader range of photon wavelengths. To fully benefit from the intrinsic layer, the PIN diode normally requires an applied reverse bias that ensures a depletion region extension all the way through this layer. Then, the region's electric field also extends through the layer to accelerate those carriers generated in a larger sensitive volume. As expected, this added layer increases the transit distance for the carriers, but the low doping of the intrinsic region retards recombination to assure that more of the generated carriers complete the transition.

1.3.2 Avalanche photodiodes

For low-light conditions and high-speed requirements, avalanche photodiodes provide significant benefits but impose difficult bias requirements. Biasing a photodiode near its breakdown voltage introduces avalanche multiplication to amplify the fundamental photodiode current. Then, even very low-light levels produce measurable signals. Further, this operation improves speed through the accelerating effect of the bias and the reduced load resistance required for a given voltage output. As described in later chapters, the magnitude of this resistance seriously affects the bandwidth of practical photodiode circuits.

Figure 1.6 illustrates the chain reaction that produces the internal gain of the avalanche photodiode. There, an incident photon begins the process by freeing a first hole-electron pair within the diode's depletion region. Then, the high electric field produced by the diode

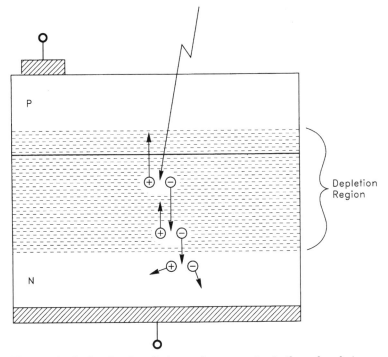

Figure 1.6 Avalanche photodiodes produce current gain through a chain reaction in which carriers released by photons and accelerated by the depletion region release further carriers through impact ionization.

bias accelerates and imparts energy to the freed carriers. In the figure, the electron of this first pair produces a second pair upon collision with another atom in the lattice, again within the depletion region. The region's electric field accelerates these new carriers to continue the multiplication process. This process continues even to areas outside the depletion region, but there, the lack of an accelerating electric field makes the regenerative nature of the process gradually decay. Through this avalanche multiplication, the photodiode current becomes $i_{pa} = Mi_{p0}$, where i_{p0} is the photodiode current produced by the same illumination but with zero diode bias. Avalanche photodiodes achieve multiplication factors as great as $M = 100$ before practical limitations preclude further increase.

An added noise effect and difficult gain control ultimately limit the usable gain of the avalanche photodiode. The impact ionization that generates the gain introduces another degree of randomness to the photodiode's current flow, and this randomness represents noise current. Making M too great degrades the signal-to-noise ratio, but for an optimum M value, the added gain supplied to the signal outweighs the added noise and improves the signal-to-noise ratio. Accurate control of this gain presents the most serious problem in practical applications. Manufacturing tolerances and temperature variations potentially produce dramatic changes in the multiplication gain M. As described, biasing the diode at the verge of breakdown produces a chain reaction, and as with any such reaction, only precise control produces the desired result. Near this breakdown point, the multiplication gain displays an almost exponential sensitivity to the applied voltage. As a result, tolerance variations in the bias voltage and in the actual diode breakdown voltage may require individual circuit adjustment. Even then, temperature variations change the breakdown voltage by around[4] 0.11 percent per °C.

One solution to these difficulties employs a dummy diode to automatically bias the primary diode at a known operating point. There, biasing an unilluminated photodiode from a known source of current establishes a compensated bias reference voltage. Matching this diode to the primary device removes all of the variables mentioned above. Then, transferring the voltage developed across the dummy to the actual photodiode assures stable gain control.[2]

1.4 Position-Sensing Photodiodes

Multiple-element and lateral photodiodes produce signals indicating the position of a light beam on the diode's surface. Two or more photodiodes on a common integrated circuit chip, as illustrated in Fig. 1.7, indicate position through the relative output signals of the

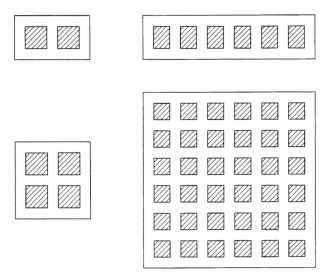

Figure 1.7 The physical locations of individual diodes in multiple-element photodiodes permit detection of a light beam's position on the photodiodes' surfaces.

individual diodes. In the simplest case, a dual photodiode produces single-axis information, and a linear diode array extends resolution for this application. Similarly, a quad photodiode, with its diodes arranged in quadrature format, provides two-axis information, and two-dimensional arrays extend resolution for this case. Lateral photodiodes also indicate position through the relative magnitudes of output signals, but these signals develop at multiple outputs on a single photodiode.[5] There, an internal current divider separates the diode current into two signals having magnitudes determined by a light beam's position. For two-axis monitoring, duo-lateral photodiodes add a second current divider and tetra-lateral photodiodes present a four-way current divider. Photodiode amplifiers described in Chap. 10 convert the resulting photodiode output currents into position-indicating voltages.

1.4.1 The basic lateral photodiode

Being a single, continuous element, the lateral photodiode provides higher resolution by avoiding the inherent quantizing error of multiple-element devices. Figure 1.8 illustrates this photodiode's structure with its characteristic output electrodes. The bottom electrode or contact covers the entire span of the diode and will collect carriers at any point along this span. However, the top electrode separates into two segments that lie at opposite sides of the diode chip. The carriers collected

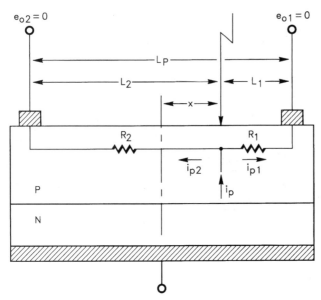

Figure 1.8 With the lateral photodiode, internal resistance forms a current divider that separates i_p into two components whose magnitudes reflect the position of an incident light beam.

there must first travel to these electrodes, and this feature results in the position indication. In the figure, a light beam incident at a given horizontal location produces a localized current. As described before, this current actually results from conduction carriers released within the body of the diode. From there, the electric field of the junction sweeps them vertically to the top and bottom of the diode in the same horizontal location as the light beam. Those swept to the top form the i_p current illustrated, and this current must travel laterally through the resistance of the semiconductor material to reach the diode's contacts. This resistance provides the current divider action to produce a linear position indication.

To quantify the result, first consider the current divider of the photodiode. There, current i_p divides into two components, i_{p1} and i_{p2}, that travel separate paths to the two top electrodes. Two circuit conditions potentially influence this current division: the resistances of the two paths, R_1 and R_2, and the signal voltages at the two output terminals, e_{o1} and e_{o2}. Linear position indication requires that $e_{o2} = e_{o1}$, so only the resistances determine the result. Usually, the photodiode amplifiers described later hold both e_{o1} and e_{o2} at 0 V and satisfy this requirement. Then, for the superimposed circuit model shown, the $e_{o1} = e_{o2}$ condition effectively shorts the output terminals together, making R_1 and R_2 the classic current divider. This produces

$i_{p1} = (R_2/R_P)i_p$ and $i_{p2} = (R_1/R_P)i_p$, where $R_P = R_2 + R_1$ is the interelectrode or positioning resistance of the photodiode.

Next, simple analysis of the dimensions illustrated translates these current divider results into the desired position indication. There, the current i_p results from a light beam incident at a distance x from the centerline of the photodiode, and the other dimensions shown relate x to the length of the diode. The three resistances of the current divider equations result from corresponding lengths, L_1, L_2, and L_P. There, L_1 and L_2 denote the path lengths followed by i_{p1} and i_{p2}, and L_P denotes the interelectrode or positioning distance, $L_P = L_1 + L_2$. A uniform, lateral resistivity ρ in the P region makes $R_1 = \rho L_1$, $R_2 = \rho L_2$, and $R_P = \rho L_P$. Substituting these expressions in the previous i_{p1} and i_{p2} equations directly replaces the resistances with lengths for $i_{p1} = (L_2/L_P)i_p$ and $i_{p2} = (L_1/L_P)i_p$. From Fig. 1.8, the desired position distance is $x = L_2 - L_1$, and simultaneous solution with the i_{p1} and i_{p2} equations yields

$$x = \frac{(i_{p1} - i_{p2})}{i_P}L_P = \frac{(i_{p1} - i_{p2})}{(i_{p1} + i_{p2})}L_P$$

Thus, relative to the center of the photodiode, the x position of the light beam equals the difference in the two signal currents divided by their sum times the interelectrode spacing of the diode. This difference divided by the sum quotient normalizes the result to automatically remove errors from light intensity variation and from diode responsivity tolerance as described in Chap. 10. That chapter also presents photodiode amplifier circuits that process the i_{p1} and i_{p2} signals to produce this quotient.

1.4.2 The lateral photodiode model

For general circuit analysis the lateral photodiode requires a modified circuit model. Above, the analysis assumes that the two top terminals of the diode reside at the same voltage, making the R_1 and R_2 resistances a simple current divider. Most photodiode amplifier circuits approximate this equal voltage condition by holding the two terminals near 0 V. This condition reduces the lateral photodiode's outputs to currents that directly relate to position. However, the i_{p1} and i_{p2} equations derived above only model the diode response for this equal-voltage condition. As will be seen in later chapters, the 0 V constraint of photodiode amplifiers remains an approximation, especially at higher frequencies. Further, some application circuits connect the output terminals to nonzero impedances where the output currents develop differing signal voltages at their output terminals.

These practical cases require a photodiode model that represents the lateral photodiode action through the R_1 and R_2 resistances rather than the i_{p1} and i_{p2} currents. Figure 1.9 presents this model with these resistances forming the two segments of a potentiometer. There, the position of the potentiometer wiper represents the position of the light beam on the photodiode. The total resistance of the potentiometer equals the diode's positioning resistance R_P, and $R_P = R_1 + R_2$. Further dimensional analysis of Fig. 1.8 defines the R_1 and R_2 resistances in terms of R_P and the relative position of the light beam. Again, assuming a uniform lateral resistivity ρ in the P material makes $R_1 = \rho L_1, R_2 = \rho L_2$, and $R_P = \rho L_P$. Simultaneous solution of these three equations produces the intermediate results $R_1 = (L_1/L_P)R_P$ and $R_2 = (L_2/L_P)R_P$. Then, noting that $L_P = L_1 + L_2$ and $x = L_2 - L_1$ reduces these equations to

$$R_1 = \frac{L_P - x}{2L_P} R_P$$

$$R_2 = \frac{L_P + x}{2L_P} R_P$$

These equations define the current dividing resistances in terms of photodiode constants R_P and L_P and the light position x. Then, including the Fig. 1.9 model in any circuit model permits an analysis that defines x in terms of the diode constants and the circuit conditions.

In practice, parasitic series resistance limits the position detection accuracy of the lateral photodiode. There, an inevitable electrode contact resistance adds to both R_1 and R_2, preventing them from reducing to their ideal zero values as x approaches either L_P or $-L_P$ above. However, adjusting the model to account for this practical limitation

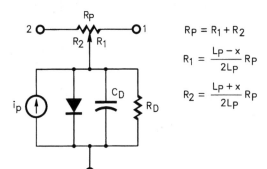

Figure 1.9 A potentiometer models the current-dividing output resistances of the lateral photodiode with R_1 and R_2 segments determined by the position x and by diode constants.

permits more accurate analysis for precision applications. For those cases

$$R_1 = \frac{L_P - x}{2L_P} R_P + R_C$$

$$R_2 = \frac{L_P + x}{2L_P} R_P + R_C$$

where R_C is the contact resistance.

1.4.3 The duo-lateral photodiode

Two alternatives to the basic lateral photodiode extend the position indication from single axis to two axis. Duo-lateral photodiodes simply utilize both the top and bottom diode surfaces in the lateral sensing but complicate fabrication and preclude diode bias. Tetra-lateral photodiodes remove these limitations but degrade response linearity and complicate signal processing. Figure 1.10 illustrates the duo-lateral structure with two pairs of electrodes contacting the diode's P and N layers. As before, the X_1 and X_2 electrodes at opposite sides of the chip collect photo current supplied to the P layer. Previously, a continuous electrode covered the diode's bottom surface to make contact with the N layer and conduct the return current. Here, the bottom electrode divides into the Y_1 and Y_2 contact strips placed at opposite ends of the chip. These contact strips now collect the diode's return current through the same lateral action as previously described for the top electrodes. However, the orthogonal orientation of the top and bottom contact pairs make the two respond to opposing dimensional axes of a light beam position.

The simplicity of the duo-lateral photodiode structure extends previous results to this case. First consider the ideal case where photodiode amplifiers hold the two electrodes of a given pair at the same voltage. Then, the two X electrodes again receive portions of the photodiode current as determined by the resistive divider presented by the P region. Similarly, the Y electrodes here receive currents apportioned by the resistivity of the N region. Then, the X and Y current components

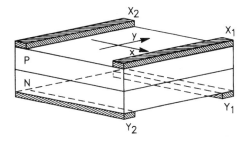

Figure 1.10 A duo-lateral photodiode makes both diode surfaces perform lateral current division with two electrode pairs placed for X and Y position indications.

define the light beam position through

$$X = \frac{(i_{px1} - i_{px2})}{(i_{px1} + i_{px2})} L_{Px}$$

$$Y = \frac{(i_{py1} - i_{py2})}{(i_{py1} + i_{py2})} L_{Py}$$

where L_{Px} and L_{Py} represent the interelectrode or positioning spacings of the two electrode pairs. Chapter 10 presents photodiode amplifier circuits that produce output voltages proportional to the above X and Y position indications.

For nonideal cases, a simple extension of the previous circuit model represents the duo-lateral photodiode for analysis. Shown in Fig. 1.11, this model repeats that of Fig. 1.9 with a second current-dividing potentiometer modeling the current division of the diode's bottom, or Y axis, outputs. As before, the position of the light beam defines the equivalent positions of the potentiometer wipers. However, the orthogonal orientation of the electrode pairs makes the wiper positions independent. Extending the single-axis analysis reflects the wiper positions through the resistances of the potentiometer segments. This produces matching sets of equations for the two axes as expressed by

$$R_{1x} = \frac{L_{Px} - x}{2L_{Px}} R_{Px} \qquad R_{2x} = \frac{L_{Px} + x}{2L_{Px}} R_{Px}$$

$$R_{1y} = \frac{L_{Py} - y}{2L_{Py}} R_{Py} \qquad R_{2y} = \frac{L_{Py} + y}{2L_{Py}} R_{Py}$$

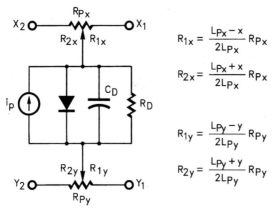

Figure 1.11 Adding a second potentiometer to the lateral photodiode model of Fig. 1.9 extends this model to the duo-lateral case.

where R_{Px} and R_{Py} represent the interelectrode resistances, and L_{Px} and L_{Py} represent the interelectrode spacings.

Two features of the duo-lateral photodiode structure detract from its otherwise simple extension of the lateral photodiode action. First, segmented electrical contacts must be made to the bottom of the diode, and conventional planar-technology integrated circuit fabrication only permits this on the top surface. Diffused vias from the top surface to the bottom layer overcome this limitation; however, these vias increase the series resistances that produce the greatest position-sensing errors. In addition, the duo-lateral structure precludes or significantly complicates the application of reverse bias to the photodiode. As described before, this bias reduces the diode capacitance to improve response speed. Normally, one side of a photodiode supplies the signal and the other side remains free to be biased from a dc source. With the duo-lateral structure, both sides of the diode supply the signal and cannot be directly connected to a bias source.

1.4.4 The tetra-lateral photodiode

With some performance compromise, tetra-lateral photodiodes remove these restrictions by placing both the X and Y electrode pairs on the top surface. Figure 1.12 illustrates this structure with four upper electrodes placed at the sides and ends of the top surface and with a continuous bottom electrode. The upper electrodes accommodate conventional fabrication practices, and the single lower one again permits diode bias. At the upper surface, resistances still divide the photodiode current into components that travel to the various output electrodes. However, two-dimensional current flow complicates the current division here. Previously, the basic lateral and duo-lateral photodiodes divided the photodiode current in a given diode layer into lateral flows to two electrodes placed along the same axis. There, simple geometric boundary conditions established by the electrode terminations

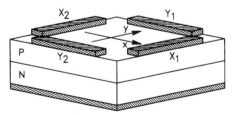

Figure 1.12 A tetra-lateral photodiode places two electrode pairs on the top surface for two-dimensional position sensing without the need for segmented bottom contacts.

produce a linear position response. Here, four electrodes on two axes complicate the boundary conditions and introduce distortion in the tetra-lateral photodiode response. Varying the diode geometry partially compensates for this distortion with pin-cushion-shaped diode areas and corner diode contact points.[6]

Figure 1.13 models the tetra-lateral photodiode with a four-way output current divider. For this divider, the X and Y coordinates of the beam position determine four resistances values in essentially the same way as described for the duo-lateral case. Thus, the corresponding resistance equations remain the same. With this model, considering two separate current divisions extends previous ideal analysis results to this case. As before, this analysis holds for equal output terminal voltages such as established by the zero voltage constraint of most photodiode amplifiers. This constraint effectively places all of the x and y resistors of the figure in parallel for simple current division. In this condition, the i_p current first divides into the net x and y components $i_x = i_{x1} + i_{x2}$ and $i_y = i_{y1} + i_{y2}$. Resistances $R_x = R_{x1} \| R_{x2}$ and $R_y = R_{y1} \| R_{y2}$ control this division and produce the current components $i_x = R_y i_p/(R_x + R_y)$ and $i_y = R_x i_p/(R_x + R_y)$. Next,

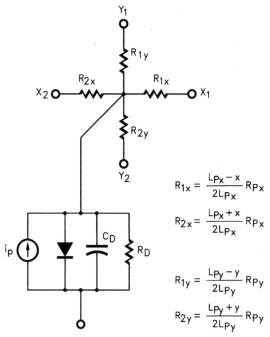

Figure 1.13 The single-surface output of the tetra-lateral photodiode forms a four-way current divider with resistances corresponding to X and Y coordinates.

the i_x and i_y components divide into their respective i_1 and i_2 subcomponents just as described before for the duo-lateral photodiode. Thus, the subcomponents identify the beam position through the same response equations and for the tetra-lateral photodiode:

$$x = \frac{(i_{px1} - i_{px2})}{(i_{px1} + i_{px2})} L_{Px}$$

$$y = \frac{(i_{py1} - i_{py2})}{(i_{py1} + i_{py2})} L_{Py}$$

Although these equations remain the same, the tetra-lateral photodiode complicates the signal processing that develops the corresponding voltage signals. In this case, the denominators of the above expressions vary with the beam's position, increasing the dynamic range requirement of that processing. Previously, the duo-lateral photodiode conducted the full diode current through separate X and Y sensing layers. This made the two denominators, $i_{x1} + i_{x2} = i_p$ and $i_{y1} + i_{y2} = -i_p$, independent of the beam's position. However, as described above, the four-way current division of the tetra-lateral photodiode makes $i_{x1} + i_{x2} = R_y i_p/(R_x + R_y)$ and $i_{y1} + i_{y2} = R_x i_p/(R_x + R_y)$, where $R_x = R_{x1} \| R_{x2}$ and $R_y = R_{y1} \| R_{y2}$. Thus, each denominator represents only a portion of the total i_p signal. Further, that portion varies with beam position through the varying resistances of the current divider. In the extreme case, a beam position very close to an electrode of one axis diverts most of i_p to that electrode and leaves only a small current to support the other axis. Then, signal levels become very small for the unsupported axis, and the subsequent signal processing must support a wider dynamic range to accommodate varied beam positions.

References

1. S. Gage, *Optoelectronics/Fiber-Optics Applications Manual*, McGraw-Hill, New York, 1977.
2. J. Wilson and J. Hawkes, *Optoelectronics*, Prentice International (U.K.), 1989.
3. S. Middelhoek and S. Audet, *Silicon Sensors*, Academic Press, New York, 1989.
4. A. Chappell, *Optoelectronics; Theory and Practice*, Texas Instruments, 1976.
5. F. Daghighian, "Optical Position Sensing with Duolateral Photodiodes," *Sensors*, November 1994, p. 31.
6. Large Area PSD Series, Hamamatsu, CR-3000, 1989.

Chapter 2

The Basic Amplifier

This chapter demonstrates the superiority of the operational amplifier (op amp) current-to-voltage converter as a photodiode amplifier through comparisons with two obvious alternatives. Both alternatives employ voltage-mode signal processing and fail to match the performance of the current-to-voltage converter. First, direct voltage monitoring of a photodiode employs familiar circuitry but results in severe nonlinearity and high dc offset. The op amp current-to-voltage converter largely removes both problems but retains a potentially significant offset for higher-gain applications. Replacing the current-to-voltage conversion resistance with a tee network controls this residual offset. A second voltage-mode alternative supplies the diode current directly to a resistive load but results in a severe bandwidth limitation. This limitation results from signal voltage swing across the diode capacitance and the accompanying capacitive shunting. Once again, switching to the current-to-voltage converter largely removes the limitation, in this case, by isolating the diode capacitance from the signal swing. Simple in structure, the current-to-voltage converter consists of just an op amp and a feedback resistor and provides dramatic performance improvements. However, when combined with a photodiode, this simple circuit displays surprising multidimensional constraints as explored in subsequent chapters.

2.1 Linearity

The output signal from a photodiode can be monitored as either a voltage or current, but current monitoring offers far better linearity, offset, and bandwidth performance. The free carriers created in the photodiode by light fundamentally constitute a current signal. However, most electronic instrumentation accommodates voltage rather than current signals, requiring a conversion. By itself, the photodiode produces this current-to-voltage conversion when presented with a high load impedance as in Fig. 2.1. There, placing the photodiode in series with an op amp input makes the diode's load the amplifier's high input impedance. This impedance does not significantly shunt the diode's current, leaving that current to circulate within the diode. Then, the carriers generated serve to expand the diode's depletion region and the corresponding voltage across the diode e_p. Feedback around the op amp establishes an amplification of e_p just as if this voltage was an input offset of the amplifier. Thus, Fig. 2.1 produces an output voltage of $e_o = -(1 + R_2/R_1)e_p$.

While appealingly simple, this voltage-mode connection produces a nonlinear response and a large dc offset. The nonlinearity results primarily from the logarithmic current-voltage characteristic of the diode junction. As described in Sec. 1.2, a photodiode operated in the voltage-output or photovoltaic mode produces a voltage described by $e_p = V_t \ln(r_\phi \phi_e / I_D)$ where ϕ_e is the flux energy of the incident light. Thus, linear changes in ϕ_e produce logarithmic changes in e_p. In addition, the flux responsivity, r_ϕ of the equation, varies with the voltage across the diode. There, a voltage gradient across the diode's junction accelerates the carriers generated, reducing recombination and increasing r_ϕ. In the photovoltaic mode shown, the photodiode's own output voltage modulates the junction voltage to further increase nonlinearity. The voltage-mode circuit also produces a large dc offset due the flow of the op amp's input bias current I_{B-} through the high resistance of the photodiode. This resistance is the very large reverse resistance of the diode and produces a dc error that adds directly to e_p.

Current-mode monitoring removes both e_p and I_{B-} from the photodiode with different degrees of success. This alternative virtually

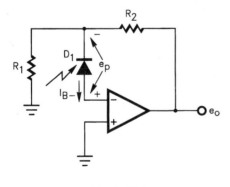

Figure 2.1 A straightforward application of a photodiode places the diode in series with an op amp input to avoid impedance loading but results in a nonlinear response and a large dc offset.

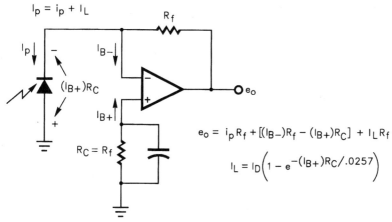

Figure 2.2 The current-to-voltage converter removes signal voltage from a photodiode to largely remove the Fig. 2.1 limitations but retains a leakage-producing voltage across the diode.

eliminates e_p, and the resulting nonlinearity, but only transfers I_{B-} to a smaller resistance.[1] Current monitoring requires that the monitor circuit present zero load impedance to the diode. Then, the monitor absorbs the diode's current without producing a voltage across the diode. An op amp circuit closely approximates this zero impedance condition with the virtual ground of its feedback input. The current-to-voltage converter of Fig. 2.2 presents this virtual ground directly to the photodiode. This figure shows the basic circuit, composed of the op amp and R_f feedback resistor, plus other common circuit elements. As described later, the added R_C reduces the offset produced by amplifier input current, and the capacitor shunts this resistor's noise.

Ignoring these added elements for the moment simplifies the initial analysis of the circuit's operation. There, feedback forces the current-to-voltage converter to accept the photodiode output current without producing any significant voltage at the circuit input. In this feedback relationship, a current I_p supplied by the photodiode initially reacts with the high input impedance of the op amp. The amplifier's open-loop gain amplifies the resulting input voltage, driving the feedback resistor R_f with output voltage e_o. This drive diverts the diode's current away from the amplifier input impedance to R_f, restoring virtually zero voltage at the op amp input. In the process, the amplifier develops an output voltage equal to the diode current times the feedback resistance, $e_o = I_p R_f$. Here, $I_p = i_p + I_L$ consists of signal and leakage components, and momentarily ignoring the offset effect of I_L makes the output signal $e_o = i_p R_f$. As described in Sec. 1.1, $i_p = r_\phi \phi_e$ and $e_o = r_\phi \phi_e R_f$ displays the desired linear signal response to light flux ϕ_e. Further, the current-to-voltage converter removes the signal swing from the diode, making r_ϕ a constant and removing the second source of nonlinearity described for the voltage-mode circuit.

2.2 Offset

In photodiode applications, the current-to-voltage converter develops a dc offset primarily through two error currents. Both the input current of the amplifier and the leakage current of the photodiode flow through the R_f feedback resistor, producing dc offset components at the circuit output. Two techniques reduce the net offset error through a compensation resistor and a tee network. The compensation resistor develops an offset component that counteracts that produced by amplifier input current. However, the compensation voltage developed increases the photodiode leakage current and potentially degrades rather than improves the net circuit offset. Replacing the R_f feedback resistor with a tee network removes this compromise by reducing the compensation voltage required. In addition, the tee alternative improves the accuracy of the compensation by permitting improved resistor matching.

2.2.1 Reducing offset with a compensation resistor

In Fig. 2.2, the R_C resistor provides the compensation routinely applied to reduce the offset effect of the op amp input current. However, photodiode applications require greater care in making the decision to add this resistor. First, consider its intended compensation function. The I_{B-} input current of the op amp flows through R_f creating an output offset component equal to $(I_{B-})R_f$. The large R_f resistance values commonly used with photodiodes make this a significant offset component that exhibits a large thermal drift due to the temperature coefficient of I_{B-}. To compensate, connecting the equal resistance, $R_C = R_f$, in series with the op amp's noninverting input produces a counteracting offset effect $-(I_{B+})R_C$. Accompanying R_C in the figure, an added capacitance bypass largely removes this resistor's noise effect. Adding R_C reduces the offset error created by amplifier input currents to the residual error developed by mismatches. Matching errors between the amplifier input currents and between the two resistors retain a reduced output offset error of $V_{OSO} = (I_{B-})R_f - (I_{B+})R_C$. This residual error typically represents a factor of 5 to 10 improvement over the initial offset and drift produced by the amplifier's input current.

However, adding R_C also develops a voltage across the photodiode, introducing a new source of offset error. The compensating voltage drop across R_C, $(I_{B+})R_C$, also drops across the photodiode and produces the diode's leakage current I_L. The classic junction equation describes this leakage through

$$I_L = I_D \left(1 - e^{-(I_{B+})R_C/V_t}\right)$$

where I_D is the reverse saturation current or dark current of the photodiode, and V_t is the thermal voltage of the semiconductor material. For silicon, $V_t = 25.7$ mV, and just 100 mV across the diode essentially develops its full-scale leakage current I_D. The resulting I_L flows through R_f to produce an output offset component equal to $I_L R_f$. Adding this component to the previous result makes the net output offset voltage $V_{OSO} = [(I_{B-})R_f - (I_{B+})R_C] + I_L R_f$. With perfect matching of R_C to R_f, the output offset reduces to $V_{OSO} = (I_{B-} - I_{B+} + I_L)R_f$ and either input or leakage currents can dominate this result.

For high-gain photodiode amplifiers, the I_L term could easily dominate offset performance. There, field-effect transistor (FET) input op amps reduce the offset produced by amplifier input current, but larger-area photodiodes potentially produce large leakage currents. These amplifiers produce input currents through the leakage currents of their input FETs. Both the amplifier FETs and the photodiode produce leakage current magnitudes in proportion to their physical areas. High-sensitivity photodiodes require much larger junction areas than the typical op amp input FET and, thus, potentially larger leakage currents. Adding R_C to compensate for the leakage of a smaller input FET can produce a greater diode leakage and increase rather than decrease the net offset error. This requires evaluation of specific application conditions before adding R_C.

2.2.2 Reducing offset error with a feedback tee

Replacing R_f with a resistor tee greatly reduces offset error through reduced resistance levels. Reduced resistances decrease the $(I_{B+})R_C$ compensation voltage impressed upon the photodiode and reduce the resistance matching error of R_C to R_f. However, the tee network also imposes a limit to this offset reduction by amplifying the op amp's input offset voltage. This compromise serves as one guide in the design of the tee alternative for a given application. Later, Chap. 7 develops another design guideline that avoids the additional noise commonly encountered with a feedback tee.

Figure 2.3 shows the R_f replacement with a tee that produces an equivalent feedback resistance but with much smaller resistance levels. Considering signal conditions first, the circuit's feedback action again accepts the photodiode current I_p at the amplifier's inverting input. To do so, the feedback signal develops a voltage of $I_p R_{fT}$ across

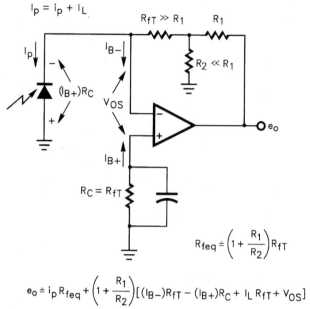

Figure 2.3 Replacing the feedback resistor with a tee network reduces offset effects through resistance-level reductions that decrease the $(I_{B+})R_C$ voltage across the photodiode and ease resistor matching.

R_{fT}. In the basic current-to-voltage converter, this would be the circuit's output voltage. However, the voltage divider formed by R_1 and R_2 here increases the output voltage e_o required to supply this $I_p R_{fT}$ voltage. This increased e_o reflects the tee's equivalent feedback resistance through $e_o = I_p R_{feq}$ where $R_{feq} = R_{fT} + R_1 + R_{fT}R_1/R_2$. In practice, R_{fT} remains by far the largest resistance of the tee and $R_{fT} \gg R_1$, making $R_{feq} \approx (1 + R_1/R_2)R_{fT}$. Thus, the tee effectively multiplies the resistance of R_{fT} by the inverse of the R_1, R_2 voltage divider ratio or $1 + R_1/R_2$. This inverse ratio characterizes the feedback action of the tee and reappears numerous times in the tee performance analyses. The tee's equivalent feedback resistance produces an output signal of $e_o \approx I_p(1 + R_1/R_2)R_{fT}$ just as if a very large feedback resistor had been used instead of the tee network. For a given current-to-voltage gain, this multiplication reduces the required resistance of R_{fT} by the same $1 + R_1/R_2$ factor.

Offset voltage analysis reveals the same resistance reduction for R_C, decreasing the compensation voltage developed across the photodiode. As before, the flow of I_{B+} in R_C offsets the op amp's noninverting input to a voltage of $-(I_{B+})R_C$. Feedback forces the amplifier's inverting input to track this offset, impressing this reduced voltage across the diode and producing a leakage current. At the input of the tee network, the amplifier's I_{B-} input current adds to the diode's leakage current. Within the tee, the combined current $I_L + I_{B-}$ flows in R_{fT}, developing a voltage that adds to the $-(I_{B+})R_C$ offset at the noninverting input. Thus, the net offset voltage at the junction of R_1 and R_2 divider becomes $(I_L + I_{B-})R_{fT} - (I_{B+})R_C$. From there, feedback action amplifies this net offset to produce an output offset voltage of $V_{OSO} = (1 + R_1/R_2)[(I_{B-})R_{fT} - (I_{B+})R_C + I_L R_{fT}]$.

Examination of this result reveals the two offset reduction benefits of the tee. As before, making $R_C = R_{fT}$ above minimizes the effect of amplifier input currents to the degree permitted by the matching of R_C to R_{fT} and I_{B-} to I_{B+}. With perfect resistor matching, the output offset reduces to $V_{OSO} = (I_{B-} - I_{B+} + I_L)(1 + R_1/R_2)R_{fT} = (I_{B-} - I_{B+} + I_L)R_{feq}$. For a given transresistance, with R_{feq} equal to the traditional R_f, this offset expression duplicates that of the basic current-to-voltage converter described before. However, in practice, the reduced resistance levels of the tee permit improved resistor matching and reduce the diode leakage to decrease the V_{OSO} actually produced. Reduced resistance improves the probability of accurate matching for the $R_C = R_{fT}$ condition that initially minimizes V_{OSO}. The basic current-to-voltage converter requires matching R_C to the typically large R_f resistance. There, very high R_f resistances make parasitic resistance shunting significantly degrade the accuracy of this match-dependent compensation. The more moderate-valued resistors of the tee reduce the significance of this shunting in the tee's $R_C = R_{fT}$ resistor match.

The preceding discussion would suggest making the tee network's R_1/R_2 ratio very large to minimize the V_{OSO} components described above. However, another offset component and noise complications impose compromises in the R_1/R_2 selection. The tee also amplifies the op amp's input offset voltage V_{OS} and input noise voltage by the factor $1 + R_1/R_2$. For offset considerations, the added V_{OS} effect increases the net output offset voltage to $V_{OSO} \approx (1 + R_1/R_2)[(I_{B-})R_{fT} - (I_{B+})R_C + I_L R_{fT} + V_{OS}]$. Noting that $R_{fT} = R_{feq}/(1 + R_1/R_2)$ and assuming that $R_C = R_{fT}$ reduces this expression to $V_{OSO} = (I_{B-} - I_{B+} + I_L)R_{feq} + (1 + R_1/R_2)V_{OS}$. Thus for a given R_{feq}, increasing R_1/R_2, to reduce R_C and I_L, simultaneously increases the offset effect of V_{OS}. Taken to the extreme, increasing R_1/R_2 makes this effect dominant and $V_{OSO} \approx (1 + R_1/R_2)V_{OS}$. Then, this new dominant component would actually increase offset in the tee conversion. An arbitrary but practical design limit avoids this increase. Limiting the V_{OS} effect to one-tenth that of the leakage current effect produces the guideline

$$\frac{R_1}{R_2} \leq \frac{I_{B-} - I_{B+} + I_L}{10 V_{OS}} R_{feq} - 1$$

As described before, I_L here also remains a function of R_1/R_2, potentially making the design solution iterative at this point. Also, a second design guideline results from the increased noise developed with a tee as developed in Chap. 7. There, the design of the feedback tee optimizes a bandwidth-versus-noise tradeoff. For a given application, balancing the offset, noise, and bandwidth constraints defines the final tee design.

2.3 Bandwidth

Junction capacitance restricts signal bandwidth for most photodiodes and often severely so. Any signal voltage developed across the diode reacts with this capacitance, shunting the diode's output current. The basic photodiode amplifier or current-to-voltage converter greatly extends bandwidth by largely isolating the photodiode from the signal voltage. This circuit holds the diode voltage near zero while supplying the signal voltage at the op amp output.

First for comparison, Fig. 2.4 shows a second voltage-mode circuit that illustrates this mode's restricted bandwidth.[1] There, signal current i_p from the photodiode flows in load resistor R_L, producing a signal voltage. The op-amp voltage follower isolates R_L from any impedance loading error at the circuit output and ideally delivers $e_o = -i_p R_L$. However, this simple circuit also develops the full signal voltage e_o across the photodiode and its capacitance. The resulting signal current in this capacitance shunts the diode current at higher frequencies, producing a bandwidth limit. Figure 2.5 models the circuit for bandwidth analysis. There, the photodiode acts basically as a current

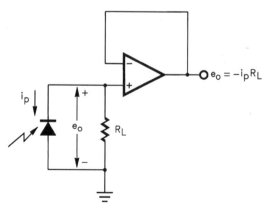

Figure 2.4 A second, straightforward application of a photodiode provides a controlled load resistance but again develops the output signal voltage across the diode.

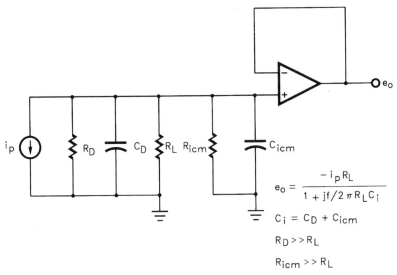

Figure 2.5 Replacing the photodiode with its fundamental elements and separating the input impedance elements from the op amp models Fig. 2.4 for bandwidth analysis.

source i_p with shunting elements R_D and C_D. R_D represents the high resistance of a zero-biased junction, and practical levels of load R_L remain small by comparison. Thus, R_D can generally be ignored. The follower amplifier shown presents additional shunting through its input resistance and capacitance, R_{icm} and C_{icm}. There, R_{icm} represents the very high, common-mode input resistance of the op amp and can also be ignored. However, the capacitive elements of both the diode and amplifier significantly shunt R_L, producing a bandwidth limit.

Thus, the circuit produces the ideal response $e_o = -i_p R_L$ only until capacitive shunting diverts i_p from R_L at higher frequencies. There, the combined capacitance of the input circuit, $C_i = C_D + C_{icm}$, produces a response roll off with a pole frequency of $f_p = 1/2\pi R_L C_i$. Then, for Fig. 2.5

$$\frac{e_o}{i_p} = \frac{-R_L}{1 + jf/f_p}$$

For this single-pole response, the circuit's −3-dB bandwidth equals the pole frequency f_p. For most photodiode applications, large values of R_L and C_D would severely restrict bandwidth.

The above equations reflect a typical gain-versus-bandwidth compromise. Increasing R_L produces greater gain but reduces the bandwidth set by f_p in direct proportion. From a circuit perspective, this

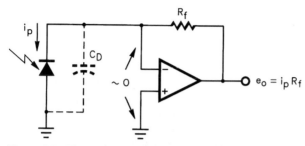

Figure 2.6 The simple current-to-voltage converter isolates the load voltage swing from the photodiode capacitance for improved bandwidth.

compromise results from impressing the signal voltage across the circuit capacitances. Increasing R_L to raise gain also increases the signal voltage across these capacitances. This increases the portion of the i_p signal current that the capacitances shunt away from the load. Then, the -3-dB response point of the circuit decreases to a lower frequency. To avoid this compromise, the signal voltage would ideally develop on the load resistor but not on the capacitances.

The current-to-voltage converter approximates this ideal as shown in Fig. 2.6. In the ideal amplifier case, this circuit reduces the signal voltage across the photodiode and its capacitance to zero. There, the feedback action of the op amp forces the two amplifier inputs to the same voltage, presenting a virtual ground to the photodiode output current. The load's signal voltage transfers to the op amp output, isolated from the diode.

In practice, the amplifier's high, but finite, open-loop gain only approximates the zero voltage condition above. The finite gain leaves part of the signal voltage across the photodiode and its capacitance. This residual signal introduces a new bandwidth limit and the potential for circuit oscillation as described in Chap. 3. Both limitations result from the feedback pole formed by R_f and C_D. In addition, the op amp's input noise voltage adds to the residual input signal and develops a noise current in C_D. That current increases with frequency and flows into R_f to produce a high-frequency gain peaking that often dominates noise performance. Chapter 5 analyzes this noise response and later chapters present means to limit the noise effect.

References

1. J. Graeme, "FET Op Amps Convert Photodiode Outputs to Usable Signals," *EDN*, October 29, 1987, p. 205.

Chapter

3

Bandwidth and Stability

Three separate response limits each potentially set the bandwidth of the current-to-voltage converter when operated as a photodiode amplifier. These result from parasitic capacitance, the limited bandwidth of the op amp, and phase compensation. The resulting pole frequencies define each response limit. A fourth limit, the frequency response of the photodiode, also potentially sets bandwidth, but generally one of the first three mentioned produces a lower frequency limit than the diode. In the simplest case, the inevitable parasitic capacitance bypasses the feedback resistor, and this effect generally sets the bandwidth for higher-gain cases. Lower-gain applications reduce this resistance and the significance of the parasitic bypass. Then, the response limits imposed by the op amp bandwidth and the phase compensation compete to set the circuit's bandwidth.

The latter two limits prove more difficult to determine. Despite the circuit's simplicity, the current-to-voltage converter can exhibit complex ac performance. As a photodiode amplifier, this circuit produces a two-pole, rather than single-pole, response, resulting from an input circuit that appears like a parallel L-C circuit. This condition occurs when parasitic capacitance does not dominate bandwidth, and the limited bandwidth of the op amp introduces these two poles in the circuit response. Then, the higher feedback resistances of photodiode applications often make this equivalent L-C circuit resonate or even oscillate. This behavior requires the phase compensation that introduces the third response limit. Making this new limit coincide with that of the op amp's bandwidth optimizes the circuit's bandwidth-versus-stability compromise. Feedback analysis develops design equations for the corresponding compensation selection through frequency response evaluation and phase margin analysis.

Some applications require greater control of gain peaking than provided by the above compromise. Making the above two limits coincident optimizes the circuit's 3-dB bandwidth but also permits a 3-dB gain peak. More conservative phase compensation reduces this peak as guided by the response characteristics of a second-order system. Translating these results for the photodiode amplifier produces an alternate design equation that permits compensation selection to limit peaking at any specified level.

The analyses of this chapter focus upon the basic photodiode amplifier but also develop a bandwidth and phase compensation background that extends to the alternative amplifiers described in later chapters. In addition, this background applies to other op amp applications that present source capacitance to the amplifier. Chapter 4 describes methods for improving bandwidth, and Chap. 7 describes methods for optimizing a later noise-versus-bandwidth compromise.

3.1 The Inherent Response Limits

Two of the three response limits described above result for any photodiode amplifier formed with a current-to-voltage converter. Any such amplifier remains subject to parasitic capacitance and limited op amp bandwidth. The capacitance produces a predictably simple response limit, and a mathematical test determines whether or not this limit controls the circuit's bandwidth. The limited op amp bandwidth leaves a residual signal across the photodiode capacitance, and the resulting capacitive current actually reacts with a nonzero impedance at the photodiode amplifier input. Still, the resulting pole and its bandwidth limit typically reside at a far higher frequency than achieved with voltage-mode monitoring.

3.1.1 The response limit of parasitic capacitance

Higher feedback resistance levels potentially make parasitic capacitance bypass dominate the photodiode amplifier's response and set the circuit bandwidth. Figure 3.1 illustrates this simple case with stray capacitance C_S bypassing feedback resistor R_f. Good construction practices limit C_S to around 0.5 pf, but the higher R_f values of photodiode amplifiers make this bypass significant. At higher frequencies, C_S shunts the i_p current away from R_f to produce a response roll off with a pole at $1/2\pi R_f C_S$. When dominant, the resulting single-pole roll off makes the −3-dB bandwidth equal this pole frequency. Otherwise, poles produced by the op amp bandwidth limit, or the circuit's phase compensation determine the overall circuit bandwidth.

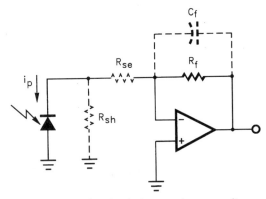

Figure 3.1 In the simplest case, stray capacitance C_S sets bandwidth for the photodiode amplifier through bypass of R_f.

To determine which case applies, design equations developed in subsequent sections of this chapter permit a simple test. Those sections develop equations prescribing the phase compensation capacitor that must be added to intentionally bypass R_f. In some cases, the inevitable C_S already provides sufficient bypass to retain frequency stability. Then, the phase compensation design equations prescribe a capacitance value less than or equal to C_S. There, C_S controls bandwidth, and the circuit requires no further phase compensation.

3.1.2 The response limit of op amp bandwidth

As described in Sec. 2.3, the current-to-voltage converter removes the bandwidth limit imposed by voltage-mode monitoring of a photodiode. There, the voltage-mode impresses the output signal across the diode's capacitance, and capacitance shunting rolls off this signal. The current-to-voltage converter largely avoids this shunting by isolating the diode from the signal swing. In the ideal case, this converter totally removes this original response limit. In practice, the converter leaves a residual signal voltage across the photodiode that imposes a higher-frequency but finite response limit. Op amp bandwidth sets this new limit through a double response pole.

The current-to-voltage converter removes signal voltage from the photodiode capacitance as shown in Fig. 3.2. There, only the op amp gain error signal e_o/A_{OL} remains across the diode and its capacitance. In the ideal amplifier case, an infinite op amp gain reduces the signal voltage across the photodiode to zero. Then, the feedback action of the op amp forces the two amplifier inputs to the same voltage, presenting a virtual ground to the photodiode output current. In practice, the

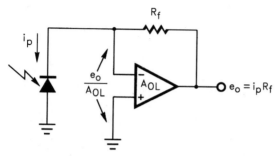

Figure 3.2 The current-to-voltage converter removes the previous voltage-mode bandwidth limit by isolating the photodiode capacitance from the output swing, but the residual gain error signal remains to produce a new limit.

amplifier's A_{OL} gain adequately approximates this ideal only at low frequencies. At higher frequencies, A_{OL} rolls off due to limited op amp bandwidth, and this produces the second response limit for the photodiode amplifier.

The resulting poles in the photodiode amplifier response define this limit, and Fig. 3.3 models the circuit of Fig. 3.2 for the corresponding analysis. To focus on the op amp limit alone, this model excludes the stray capacitance C_S described earlier. In Fig. 3.3, a current source and capacitance C_D replace the photodiode, and two other capacitors represent the input capacitance elements of the op amp. The remainder of the model presents the input circuit with an effective load re-

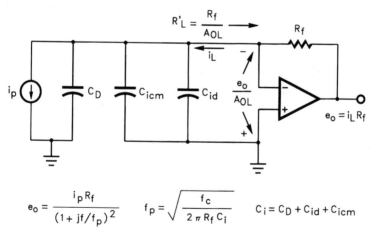

Figure 3.3 Modeling the Fig. 3.2 photodiode amplifier with an effective load resistance R'_L permits easy comparison of bandwidth limits with the voltage-mode circuit of Sec. 2.3.

sistance R'_L. This latter representation permits easy comparison with an earlier voltage-mode circuit. Previously, Sec. 2.3 described the bandwidth restriction of a photodiode that voltage-mode drives a load resistance R_L. Several comparisons in this chapter demonstrate the bandwidth improvement potentially afforded by the current-to-voltage converter.

With the Fig. 3.3 model, the break frequency of the capacitances with R'_L controls the frequency response for the current-to-voltage converter. By definition, R'_L equals the voltage developed across this resistance divided by the current i_L supplied to it. The relevant voltage equals that from the op amp inverting input to ground or simply the amplifier's gain-error signal e_o/A_{OL}. Finite open-loop gain requires this signal between the amplifier inputs to support the output voltage e_o. Current i_L produces an output voltage of $e_o = i_L R_f$, making the voltage across R'_L equal to $e_o/A_{OL} = i_L R_f/A_{OL}$. Dividing this voltage by the current i_L defines $R'_L = R_f/A_{OL}$. This resistance breaks with the circuit capacitances at

$$f_{pf} = \frac{A_{OL}}{2\pi R_f C_i}$$

where $C_i = C_D + C_{id} + C_{icm}$.

Comparison of the above result with the $f_p = 1/2\pi R_L C_i$ of the previous voltage-mode circuit provides an indication of the response improvement potential of the current-to-voltage converter. For equal gains in the two cases $R_f = R_L$, making $f_{pf} = A_{OL}/2\pi R_L C_i \approx A_{OL} f_p$. Thus, the Fig. 3.2 circuit increases the pole frequency by a factor approximately equal to the amplifier open-loop gain A_{OL}. This remains approximate because input capacitance C_i differs somewhat for the two cases. Previously, $C_i = C_D + C_{icm}$ resulted from the diode capacitance in parallel with the C_{icm} presented by a voltage follower. In this case, the amplifier adds its differential input capacitance C_{id} to the total, making $C_i = C_D + C_{icm} + C_{id}$. Very frequently, $C_D \gg C_{id}$, and the C_i difference between the two circuits becomes negligible.

Still, the A_{OL} gain varies with frequency, requiring further analysis to define the actual improvement factor. The relevant value of A_{OL} equals the amplifier's actual open-loop gain at the frequency where the f_{pf} pole occurs. To find this gain, consider an approximation to the open-loop response of the op amp. In all practical cases, f_{pf} occurs where the gain of the amplifier rolls off in a single-pole response. There, the amplifier gain magnitude becomes $A_{OL} = f_c/f$, where f_c is the unity-gain crossover frequency of the amplifier. At the frequency $f_{pf}, A_{OL} = f_c/f_{pf}$, and substituting this for A_{OL} in the above

f_{pf} equation yields a Fig. 3.3 pole location of

$$f_{\text{pf}} = \sqrt{\frac{f_c}{2\pi R_f C_i}}$$

where $C_i = C_D + C_{\text{id}} + C_{\text{icm}}$.

Once again, for comparison with the previous $f_p = 1/2\pi R_L C_i$, assume equal gains for the two circuits $R_f = R_L$ and neglect the C_i difference. Then, $f_{\text{pf}} = \sqrt{f_c/2\pi R_L C_i} \approx \sqrt{f_p f_c}$, and the f_{pf} pole location becomes the geometric mean of the old pole frequency f_p and the op amp crossover frequency f_c. As long as $f_c > f_p$, $f_{\text{pf}} \approx \sqrt{f_p f_c} > f_p$ and the current-to-voltage converter increases the pole frequency. This typically increases the frequency by a factor of 10 to 100 as can be seen by evaluating the ratio $f_{\text{pf}}/f_p \approx \sqrt{f_c/f_p}$ for typical case conditions. In the rare case where $f_c < f_p$, the current-to-voltage converter reduces rather than increases the pole frequency. Even then, however, the current-to-voltage converter still provides the improved response linearity described in Sec. 2.1.

3.2 The Phase Compensation Requirement

The above pole frequency comparison does not reflect the exact response improvement produced by the current-to-voltage converter. Further analysis shows f_{pf} above to be a double rather than single pole, and the resulting response peaking makes the -3-dB bandwidth somewhat greater than f_{pf}. However, this double pole also introduces a phase compensation requirement, and that compensation potentially reduces the circuit response. Thus, the phase compensation requirement must be satisfied before defining the final response improvement. In this section, an *L-C* tank equivalent of the photodiode amplifier first demonstrates this requirement with an intuitive analysis. There, the photodiode capacitance reacts with an inductive amplifier input impedance, and degeneration or phase compensation controls the associated resonance. Next, graphical feedback analysis confirms the resonance and establishes the basis for the phase compensation analysis of the next section. Following that, a final comparison defines the actual bandwidth improvement provided by the current-to-voltage converter in photodiode applications.

3.2.1 The *L-C* tank equivalent and phase compensation

Section 3.1.2 defines an op amp pole frequency f_{pf} based upon a resistive load $R'_L = R_L/A_{\text{OL}}$. The resulting equation accurately defines the f_{pf} frequency but not the total nature of the associated response.

The A_{OL} term of $R'_L = R_L/A_{OL}$ varies with frequency, making this load an impedance, Z'_L, rather than a resistance. This difference produces an effective L-C tank at the circuit input and requires phase compensation to preserve stability.

As frequency increases, gain A_{OL} declines, making $Z'_L = R_L/A_{OL}$ rise like an inductive impedance. The phase shift of gain A_{OL} completes this inductive character. Intuitive evaluation of the typical op amp open-loop response and a simple analysis reveal this phase characteristic. From an intuitive perspective, open-loop gain A_{OL} commonly displays a single-pole roll off throughout most of the amplifier's useful frequency range. This roll off corresponds to a phase lag of 90°. The 180° phase inversion of the basic amplifier feedback converts this lag to the 90° phase lead of an inductive impedance. Analysis quantifies this effect by including phase information in the previous approximation for A_{OL}. As described before, an op amp's single-pole roll off makes the gain magnitude $A_{OL} = f_c/f$. Switching to the s plane representation, $A_{OL} = \omega_c/s$ represents this dominant single-pole region. This makes the impedance $Z'_L = R_f/A_{OL} = R_f s/2\pi f_c$. With an s term in the numerator, this impedance appears inductive by rising with frequency and displaying a 90° phase lead.

The equivalent inductance of Z'_L, $L'_L = R_f/2\pi f_c$, forms a resonant L-C tank with C_i at the input of Fig. 3.3. Resonance occurs at the frequency $f_r = 1/2\pi \sqrt{L'_L C_i} = \sqrt{f_c/2\pi R_f C_i}$. Note that this is the same frequency previously determined for the pole location and $f_r = f_{pf}$. If the resonance occurs at a low enough frequency, it potentially produces oscillation in the current-to-voltage converter. Without phase compensation, oscillation results whenever the amplifier open-loop gain remains above unity at the resonant pole frequency f_{pf}. Beyond the unity-gain crossover frequency, the amplifier lacks the gain needed to sustain oscillation. For most photodiode amplifiers, $f_{pf} < f_c$, fulfilling this oscillation requirement. However, for higher-gain photodiode amplifiers, parasitic capacitance sometimes provides automatic phase compensation to prevent oscillation as described in Sec. 3.1.1.

Otherwise, preventing the oscillation of this L-C tank circuit requires degeneration of the tank. Figure 3.4 illustrates three degeneration or phase compensation alternatives for the photodiode amplifier. Adding the series resistor R_{se} would degenerate the circuit's tank by placing resistance between the capacitance of the diode and the inductance of the amplifier input. However, this would also place R_{se} in the input path of the photodiode current and develop a signal voltage there. That signal would also fall across the photodiode, degrading bandwidth and linearity as described in Chap. 2. Alternately, adding the shunt resistor R_{sh} across the tank would provide degeneration. However, that would also increase the noise gain of the circuit,

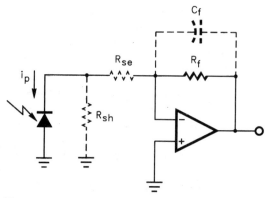

Figure 3.4 Three possible degeneration methods detune the L-C tank equivalent of the input circuit, but only C_f does so without degrading other performance.

degrading both offset and noise. Instead, adding a feedback capacitor degenerates the inductive nature of $Z'_L = R_f s/2\pi f_c$. Adding C_f in parallel with R_f makes the feedback impedance $R_f/(1 + R_f C_f s)$, and substituting this expression for R_f in the last Z'_L equation produces $Z'_L = R_f s/2\pi f_c (1 + R_f C_f s)$. Now an s term appears in the denominator of the Z'_L impedance equation, and this term counteracts the s term of the numerator. This degenerates the L-C tank and phase compensates the circuit.

3.2.2 Feedback analysis of the basic circuit

The above L-C tank analogy provides insight into the frequency stability conditions of the photodiode amplifier. However, effective control of those conditions requires a more quantitative evaluation as provided by feedback analysis. There, a combined graphical representation of the feedback demand for gain and the available gain of the op amp demonstrate these conditions for intuitive guidance of the feedback analysis.[1]

Feedback factor underlies a feedback circuit's gain demand. By definition, feedback factor is the fraction of the amplifier output signal fed back to the amplifier input. The voltage divider action of the feedback network defines this factor through the divider ratio.[2] For the photodiode amplifier, Fig. 3.5 illustrates this voltage divider with a simplified version of Fig. 3.3. Here, capacitance C_i combines the three capacitances of the input circuit to more clearly identify the feedback voltage divider. That divider, formed by R_f with C_i, introduces a low-pass filter in the feedback path and produces the feedback factor $\beta = X_{Ci}/(R_f + X_{Ci}) = 1/(1 + R_f C_i s)$. This low-pass response attenuates the feedback signal that supplies the amplifier input, placing a greater demand upon amplifier gain.

Analysis reveals this increasing demand through the circuit's response equation

$$e_o = \frac{i_p R_f}{1 + 1/A_{OL}\beta}$$

This response delivers the ideal output signal, $e_o = i_p R_f$, as long as the loop gain, $A_{OL}\beta$, remains much greater than unity. At higher frequencies, the declining β of the photodiode amplifier places increasing demands upon A_{OL} to maintain this ideal condition. However, A_{OL} also declines with increasing frequency, and the dual declines accelerate the roll off of the actual circuit response.

Graphical analysis illustrates this gain supply and demand contest and its stability consequences in Fig. 3.6. There, plotting the op amp open-loop gain along with the feedback demand for that gain makes the net conditions of the feedback loop apparent. The A_{OL} curve represents the amplifier's available gain, and the reciprocal of the feedback factor $1/\beta$ represents the gain demand of the feedback. Superimposed on this plot, the I-to-V gain curve displays the corresponding signal response of the current-to-voltage converter. As expected from the L-C tank discussion, the I-to-V curve displays a resonant peak at the frequency $f_r = f_{pf}$. Following the resonance, the I-to-V response roll off results due to a feedback demand that exceeds the available amplifier gain.

The amplifier meets this demand within the limit of the open-loop gain response of the op amp. To quantify this, the Fig. 3.5 result $\beta = 1/(1 + R_f C_i s)$ yields $1/\beta = 1 + R_f C_i s = 1 + s/2\pi f_{zf}$ where

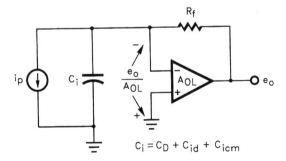

$$\beta = X_{C_i}/(R_f + X_{C_i}) = 1/(1 + R_f C_i s)$$

Figure 3.5 Simplifying the Fig. 3.3 model illustrates the R_f, C_i voltage divider that controls the circuit's feedback factor β.

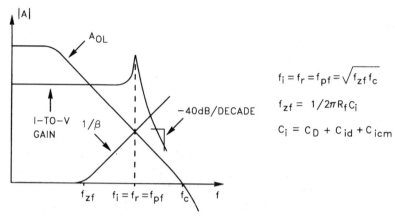

Figure 3.6 The current-to-voltage converter's A_{OL} and $1/\beta$ curves intersect at f_i, defining the frequency of the circuit's response poles and the I-to-V resonance.

$f_{zf} = 1/2\pi R_f C_i$. Here, the pole of the feedback factor becomes a zero in the $1/\beta$ feedback demand. At low frequencies in Fig. 3.6, the $1/\beta$ curve begins flat at unity as expected from simple resistor feedback around an op amp. At frequency f_{zf}, the $1/\beta$ curve begins to rise but initially remains below the open-loop gain, or A_{OL} curve. In this region, the vertical distance separating the A_{OL} and $1/\beta$ curves represents the excess gain available, or the loop gain. As the $1/\beta$ curve rises the A_{OL} curve simultaneously drops and both effects decrease the loop gain. Where the two curves cross, the gain demand reaches the limit imposed by the available gain. At frequencies beyond this crossing, the amplifier lacks sufficient gain to support the feedback demand, and the circuit's I-to-V response rolls off. This crossing, or intercept, of the A_{OL} response by the $1/\beta$ curve represents the critical intercept fundamental to feedback stability analysis.

In defining the response roll off point, the intercept at f_i also identifies the frequency of the circuit response poles. Previously, the Sec. 3.1.2 analysis defined this pole location at $f_{pf} = \sqrt{f_c/2\pi R_f C_1}$, and a two-step analysis shows that $f_i = f_{pf}$. First, the analysis above defined the zero of the $1/\beta$ response at $f_{zf} = 1/2\pi R_f C_i$. Combining this f_{zf} expression with that of f_{pf} produces $f_{pf} = \sqrt{f_{zf} f_c}$. Next, graphical evaluation of the Fig. 3.6 curves shows that f_i equals this same quantity. In the figure, the $1/\beta$ and A_{OL} curves form an approximate triangle with the horizontal axis. The two sides of the triangle have equal but opposite slopes since one follows the response of a single-zero rise and the other, of a single-pole roll off. These slope conditions produce an isosceles triangle. Then, the triangle peak at f_i occurs over the center of the triangle base. Thus, f_i occurs at the average of the base endpoints f_{zf} and f_c. Given the logarithmic nature of the frequency axis, this becomes

$$\text{Log } f_i = \frac{\text{Log } f_{zf} + \text{Log } f_c}{2}$$

Solving this expression for f_i defines that frequency as the simple geometric mean of the two endpoint frequencies,

$$f_i = \sqrt{f_{zf} f_c}$$

where $f_{zf} = 1/2\pi R_f C_i$, $Ci = (C_D + C_{id} + C_{icm})$, and f_c is the unity-gain bandwidth of the op amp. The analysis above also defined $f_{pf} = \sqrt{f_{zf} f_c}$, so $f_i = f_{pf}$.

Further evaluation of the Fig. 3.6 response curves reveals the oscillation condition that underlies the need for phase compensation. There, the slopes of the $1/\beta$ and A_{OL} curves predict frequency stability conditions through the rate of closure of the two curves. Rate of closure equals the difference between the slopes of the two curves at their intercept. Oscillation can result where the rate of closure reaches 40 dB/decade. Each 20 dB/decade of slope corresponds to 90° of phase shift, making the 40 dB/decade of the criteria correspond to 180°. Then, the op amp's gain inversion, or negative feedback, adds another 180° phase shift for a net 360° of feedback phase shift. At the f_i intercept, a 360° feedback phase shift produces oscillation. There, the A_{OL} and $1/\beta$ curves occupy the same point, making $A_{OL} = 1/\beta$ for a loop gain of $A_{OL}\beta = 1$. A feedback loop with unity gain and 360° of phase shift becomes self-sustaining, i.e., oscillation.

For Fig. 3.6, both the $1/\beta$ rise and the op amp roll off result from a single zero or pole, so both curves follow 20-dB/decade slopes. Their opposite slope polarities produce a slope difference at f_i that equals the critical 40 dB/decade, as anticipated from the earlier resonance discussion. The I-to-V response curve of the figure reflects this resonance with a high, sharp peak at f_i where oscillation will likely result. Following the peak, the I-to-V response rolls off with a -40-dB/decade slope due to the double pole at f_i. Variances in these conditions may prevent outright oscillation but still produce poor relative stability, displaying excessive overshoot and ringing. Such stability problems are familiar to everyone who has used high-feedback resistances with op amps. There, the amplifier input capacitance alone disturbs circuit response due to the same phenomena described here for the current-to-voltage converter. With photodiode amplifiers, the need for phase compensation should always be evaluated through the C_f design equations developed below.

3.3 Phase Compensation

Previously, the Sec. 3.2.1 discussion showed that the addition of a C_f feedback bypass detuned the photodiode amplifier's potential resonance. Feedback analysis quantifies the effect of the C_f degeneration and guides the selection of this capacitor through design equations. From a feedback perspective, C_f provides phase compensation as illustrated by both circuit and graphical analyses of amplifier and feedback responses. These analyses again show that C_f removes the resonance, and they quantify the value of C_f that provides optimum compensation.

3.3.1 The basic compensation

To restore frequency stability, phase compensation capacitor C_f bypasses feedback resistor R_f as shown in Fig. 3.7. This circuit repeats the earlier Fig. 3.5 and adds the C_f compensation described in Sec. 3.2.1. For high-gain applications, parasitic capacitance often provides this compensation automatically as described in Sec. 3.1.1. For other cases, the feedback bypass must generally be added to photodiode amplifiers. At high frequencies, this bypass boosts the feedback signal delivered to the amplifier input, counteracting the feedback attenuation introduced by C_i. The corresponding response zero produced by C_f also adds phase lead to reduce the net feedback phase shift. Voltage divider analysis of the Fig. 3.7 feedback network defines the

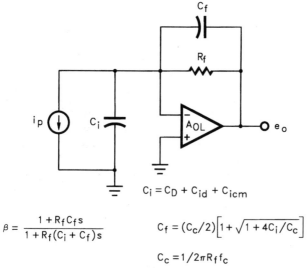

$$\beta = \frac{1 + R_f C_f s}{1 + R_f (C_i + C_f) s} \qquad C_f = (C_c/2)\left[1 + \sqrt{1 + 4C_i/C_c}\right]$$

$$C_c = 1/2\pi R_f f_c$$

Figure 3.7 Phase compensation capacitor C_f adds a zero to the circuit's feedback factor to counteract the pole produced by C_i.

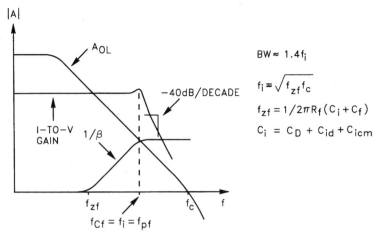

Figure 3.8 The C_f compensation of Fig. 3.7 rolls off the $1/\beta$ curve, reducing its rate of closure with A_{OL} and restoring stability.

compensated feedback factor as

$$\beta = \frac{1 + R_f C_f s}{1 + R_f (C_i + C_f) s}$$

where $C_i = (C_D + C_{id} + C_{icm})$. Here, C_f produces a zero at $f_{Cf} = 1/2\pi R_f C_f$ that counteracts the pole produced by C_i at $f_{zf} = 1/2\pi R_f (C_i + C_f)$. Note that C_f also contributes to the f_{zf} pole, but for most photodiode amplifiers, practical conditions minimize this effect with $C_f \ll C_i$. Later, a refinement to the C_f design equation accommodates the other cases.

The response zero added to β above produces a pole for the inverse function, leveling off the $1/\beta$ curve as shown in Fig. 3.8. There, the reduced $1/\beta$ slope decreases the rate of closure for improved stability, but a compromise limits the degree of improvement attainable. Choice of a large C_f could easily produce zero slope for $1/\beta$ at the f_i intercept. This would make the rate of closure a simple 20 dB/decade for 90° of phase shift and uncompromised stability. However, this choice would unnecessarily limit the circuit bandwidth. While the bypass action of C_f counteracts a feedback pole, this bypass also produces a pole for the intended circuit function. Current-to-voltage conversion depends upon the voltage developed by R_f to produce an output signal. Bypassing that resistor to reestablish stability also shunts the output signal and limits bandwidth.

3.3.2 Phase margin analysis

To optimize the -3-dB bandwidth, a compromise provides 45° of phase margin through a simple guideline that applies to all practical circuits.

Here, phase margin replaces rate of closure as the stability criteria and provides greater resolution of the relative stability condition. The phase margin reflects this condition through the net feedback phase shift at the intercept of the $1/\beta$ and A_{OL} curves. This margin equals the difference between the critical 360°, which produces oscillation, and the actual phase shift of the feedback loop at f_i. Mathematically expressed, $\phi_m = 360° + \phi_L$ for $f = f_i$. The plus sign here might seem incorrect for producing a difference; however, this remains consistent with the sign conventions associated with phase lead and phase lag. Phase lag produces a negative value of ϕ_L, and the associated minus sign there results in a subtraction in $\phi_m = 360° + \phi_L$. Also, for amplifier circuits, phase lags eventually dominate the feedback phase effects, and their negative values decrease ϕ_m.

For an op amp circuit, loop phase shift ϕ_L consists of three components easily identified by examining Fig. 3.9. There, the feedback signal travels from the output through the feedback network to the amplifier's inverting input. From there it travels through the op amp's gain response A_{OL} to the output, completing the loop. In this path, the feedback network first introduces a phase shift of ϕ_β, which represents the phase shift of the feedback factor β. The voltage divider action that defines β also determines the phase shift introduced by the feedback network. Next, the amplifier introduces the phase shift of its open-loop response ϕ_A. A third phase component results from the feedback connection to the inverting input of the op amp. This connection assures negative rather than positive feedback, and the associated gain inversion introduces a 180° phase shift. Combined, the three components of loop phase shift make $\phi_L = -180° + \phi_A + \phi_\beta$.

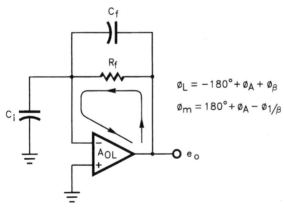

Figure 3.9 The feedback loop of a photodiode amplifier introduces phase shifts through the op amp's negative feedback, the A_{OL} response, and the feedback network.

Bandwidth and Stability   45

Substituting this expression in the previous ϕ_m result produces the phase margin result $\phi_m = 180° + \phi_A + \phi_\beta$ for $f = f_i$.

For phase margin evaluation, the response curves used in feedback analysis provide intuitive phase information directly for ϕ_A above and indirectly for ϕ_β. In Fig. 3.8, the slope of the A_{OL} response curve corresponds to ϕ_A phase shift values. No equivalent curve appears for ϕ_β, but the $1/\beta$ curve contains the information needed. Taking the reciprocal of β to express $1/\beta$ simply reverses the numerator and denominator of the β response expression. This reversal converts poles to zeroes and vice versa. Similarly, this converts phase leads to lags and vice versa. Thus, $\phi_{1/\beta} = -\phi_\beta$, and the phase margin equation becomes $\phi_m = 180° + \phi_A - \phi_{1/\beta}$ for $f = f_i$. With this equation, the slopes of feedback analysis curves communicate relative information for both phase terms needed for phase margin evaluation.

3.3.3 Phase margin analysis of the typical case

For the current-to-voltage converter, two phase margin analyses yield one C_f design equation that holds for the two circuit cases commonly found. The two differ in the proximity of intercept f_i relative to the other circuit response singularities. In the simplest and more common case, f_i occurs more than a decade in frequency above or below most of the other response poles or zeroes of the circuit. Then, such poles and zeroes produce either a full 90° or 0° of phase shift at the intercept frequency. This simplifies the phase margin analysis for a more intuitive understanding of the multiple phase effects involved. A second analysis extends the results of the first to the case where f_i resides closer to other singularities.

Figure 3.10 illustrates the simpler case with the uncompensated and compensated $1/\beta$ curves of a typical photodiode amplifier. There, the first pole of the A_{OL} response occurs well below f_i and develops a full 90° of phase lag at f_i. In contrast, the second A_{OL} pole around f_c occurs well above f_i and develops virtually no phase shift at f_i. Thus, the A_{OL} contribution to phase margin is $\phi_A = -90°$, and the preceding ϕ_m expression reduces to $\phi_m = 90° - \phi_{1/\beta}$ for $f = f_i$. In the typical case, the f_{zf} zero of the $1/\beta$ response occurs well below f_i and, when uncompensated, f_{zf} develops its full 90° of phase lead at the intercept. Thus, for the uncompensated case, $\phi_{1/\beta} = 90°$ at f_i and $\phi_m = 0°$, predicting oscillation.

To restore phase margin, phase compensation reduces the $\phi_{1/\beta}$ phase lead produced by the $1/\beta$ response. For the compensated case shown, the reduced slope of this response at the f_i intercept indicates this

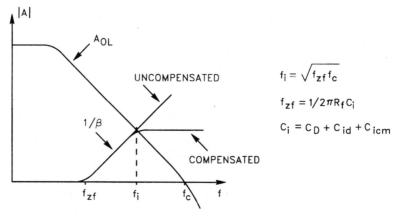

Figure 3.10 For the typical photodiode amplifier, the f_i intercept occurs at a frequency well removed from the two poles of A_{OL} and from the f_{zf} zero of $1/\beta$.

phase reduction. Phase compensation capacitor C_f produces this reduced slope through a $1/\beta$ response pole at $f_{Cf} = 1/2\pi R_f C_f$. As discussed before, a stability-versus-bandwidth compromise guides the frequency placement of this pole. Selecting f_{Cf} to produce a 45° phase margin provides good stability and optimizes the 3-dB bandwidth. For this condition, $\phi_m = 90° - \phi_{1/\beta} = 45°$, producing the phase compensation requirement $\phi_{1/\beta} = 45°$ at f_i. For this compensated case, two response singularities contribute to $\phi_{1/\beta}$, the zero at f_{zf} and the pole at f_{Cf}, making the compensation requirement $\phi_{1/\beta} = \phi_{zf} + \phi_{Cf} = 45°$ at f_i. From before, f_{zf} occurs well before the f_i intercept and develops its full 90° phase lead at f_i for $\phi_{1/\beta} = 90° + \phi_{Cf} = 45°$ which reduces the compensation requirement to $\phi_{Cf} = -45°$. The expression $\phi_p = -\arctan(f/f_p)$ describes the frequency dependence of the phase produced by a pole, and applying this to ϕ_{Cf} at $f = f_i$ produces $\phi_{Cf} = -\arctan(f_i/f_{Cf}) = -45°$. Then, solving for f_{Cf} simply produces $f_{Cf} = f_i$. Thus, choosing C_f to break with R_f right at f_i produces the desired 45° phase margin. This $f_{Cf} = f_i$ result defines the required condition but does not immediately yield a C_f design equation. Adding C_f changes f_i slightly, moving the target. Later analysis removes this interactive effect.

Note that the $f_{Cf} = f_i$ condition also makes the circuit's two bandwidth limits coincident. As described before, intercept frequency f_i marks the inevitable response roll off produced where feedback demand for gain, $1/\beta$, exceeds the available A_{OL}. Adding compensation capacitor C_f introduces an additional roll off at f_{Cf} through the C_f bypass of R_f. Making $f_{Cf} = f_i$ prevents either roll off from dominating and suboptimizing the resulting circuit bandwidth.

3.3.4 Phase margin analysis of the universal case

A continuation of the present analysis reconfirms the $f_{Cf} = f_i$ condition for a second analysis case. In the first case above, an approximation simplifies the analysis by assuming that f_i remains at least a decade removed from f_{zf} and f_c. This approximation fails with smaller photodiode capacitances, requiring a more exact analysis. This analysis begins with the assumption that $f_{Cf} = f_i$ and shows that this condition always produces $\phi_m = 45°$. In Fig. 3.11, smaller photodiode capacitances move the f_{zf} zero of $1/\beta$ to f'_{zf}. A dashed curve portrays the resulting $1/\beta$ response, and as seen, the f_{zf} change also moves the f_i intercept to f'_i. Now, both f'_{zf} and f_c reside closer to the f'_i intercept, requiring adjustment to the previous phase margin analysis. First, feedback phase shift at the new f'_i enters the range of influence of other poles that typically occur in the vicinity of f_c. There, the phase contributions of these poles increase from the simple 0° of the first analysis above. However, the f_{zf} to f'_{zf} shift also places the $1/\beta$ zero closer to the intercept, reducing its phase contribution. Then, the zero no longer develops the full 90° described for the first analysis case. Thus, the increased influence of the f_c poles and the reduced influence of the f_{zf} zero produce counteracting effects.

For first-order analyses, these two phase changes cancel, leaving the optimum choice of C_f unchanged from that of the first analysis. To demonstrate this, first consider the op amp response to be essentially two-pole in nature with the second pole occurring at the unity-gain crossover f_c. This approximation produces a phase shift of 135°

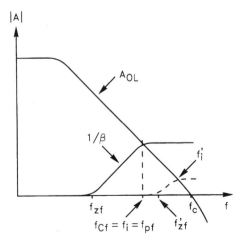

Figure 3.11 Smaller photodiode capacitances move the $1/\beta$ curve to the right, producing an f'_i that resides close to f_c and f'_{zf}.

at f_c and conservatively models typical performance. Beyond f_c, the exact phase response of the amplifier generally remains insignificant to ac performance. There the loop gain drops below the unity level needed to support oscillation or to significantly affect the transient response.

With this two-pole amplifier model, two adjustments to the previous analysis adapt the phase margin equation to this case. From before, $\phi_m = 180° + \phi_A - \phi_{1/\beta}$, for $f = f_i$, expresses the phase margin of an op amp circuit. Here, ϕ_A results from the amplifier's A_{OL} response, and $\phi_{1/\beta}$ results from the $1/\beta$ response. Previously, a single-pole A_{OL} roll off made $\phi_A = -90°$, but here the added influence of f_c increases the amplifier phase shift. Then, $\phi_A = -90° + \phi_c$ where ϕ_c represents the phase effect of the second pole modeled at f_c. This change makes the phase margin equation $\phi_m = 90° + \phi_c - \phi_{1/\beta}$ for $f = f_i$. Also previously, a single-zero $1/\beta$ rise followed by a pole at f_i produced $\phi_{1/\beta} = 90° - 45°$. There, the f_{zf} zero produced the 90° and the f_{Cf} pole produced the $-45°$. Here, the zero fails to produce its full 90° phase shift, and $\phi_{1/\beta}$ becomes $\phi_{1/\beta} = \phi_{zf} - 45°$, where ϕ_{zf} is the phase lead produced by the zero at f_{zf}. Previously, the $-45°$ contribution of f_{Cf} resulted from placing this pole at the f_i intercept. Retaining this $-45°$ value continues the $f_{Cf} = f_i$ condition. Substituting $\phi_{1/\beta} = \phi_{zf} - 45°$ in the last ϕ_m equation yields $\phi_m = 135° + \phi_c - \phi_{zf}$ for $f = f_i$.

Replacing ϕ_c and $\phi_{\beta zf}$ with their frequency response functions confirms the 45° phase margin for this second analysis case. From before, $\phi_p = -\arctan(f/f_p)$ describes the phase response of a pole, and applying this to ϕ_c at $f = f_i$ produces $\phi_c = -\arctan(f_i/f_c)$. Similarly, $\phi_z = \arctan(f/f_z)$ describes the phase response of a zero, and applying this to ϕ_{zf} at $f = f_i$ produces $\phi_{zf} = \arctan(f_i/f_{zf})$. Substituting these two results in the above ϕ_m expression yields $\phi_m = 135° - \arctan(f_i/f_c) - \arctan(f_i/f_{zf})$. From before, $f_i = \sqrt{f_{zf}f_c}$, and substitution of this expression in the last equation produces

$$\phi_m = 135° - \arctan\sqrt{\frac{f_c}{f_{zf}}} - \arctan\sqrt{\frac{f_{zf}}{f_c}}$$

The frequency variables of the above equation reside in the form $\arctan(a/b) + \arctan(b/a)$, and trigonometric analysis shows this combination to always equal 90°. Then, $\phi_m = 135° - 90° = 45°$. Thus, independent of the location of f_{zf} and f_i the phase margin remains

$$\phi_m = 45° \quad \text{for } f_{Cf} = f_i$$

where $f_i = \sqrt{f_{zf}f_c}$, $f_{zf} = 1/2\pi R_f(C_i + C_f)$, and $C_i = C_D + C_{id} + C_{icm}$.

3.3.5 Selecting the phase compensation capacitor

Analysis of the $f_{Cf} = f_i$ condition yields two design equations through approximate and exact solutions. The approximate solution serves most cases and yields an easily remembered design equation. With $f_{Cf} = f_i$ and $f_{Cf} = 1/2\pi R_f C_f$, $C_f = 1/2\pi R_f f_i$. Then, the previous $f_i = \sqrt{f_{zf} f_c}$ produces $C_f = 1/2\pi R_f \sqrt{f_{zf} f_c}$, where $f_{zf} = 1/2\pi R_f (C_i + C_f)$. However, an equation interaction prevents direct use of this result. Capacitance C_f depends on f_{zf}, which in turn depends on C_f. This occurs because the addition of the C_f phase compensation moves the f_{zf} zero of the $1/\beta$ rise in Fig. 3.11. Then, $f_i = \sqrt{f_{zf} f_c}$ follows, moving the target of the compensation selection. Two approaches remove this interaction either through approximation or simultaneous equation solution. In the simpler case, larger photodiodes make $C_D \gg C_f$ for $C_i \gg C_f$ and $f_{zf} \cong 1/2\pi R_f C_i$. Substituting this approximate f_{zf} in $C_f = 1/2\pi R_f \sqrt{f_{zf} f_c}$ yields the approximate design equation

$$C_f = \sqrt{\frac{C_i}{2\pi R_f f_c}} \quad \text{for } \phi_m = 45° \text{ when } C_i \gg C_f.$$

From before, $C_i = C_D + C_{id} + C_{icm}$. Note that, at the beginning of the C_f design process, C_i is known but C_f is not, and the $C_i \gg C_f$ test cannot be performed. However, simply calculating C_f by the above equation and then performing the test determines the equation's validity for a given case. If this test fails, simply use the more exact equation derived later.

The above result simplifies to an easily memorized relationship in which C_f becomes the geometric mean of two circuit capacitances. This requires defining an artificial capacitance, $C_c = 1/2\pi R_f f_c$, which relates R_f to f_c just as R_f relates to C_i in $C_i = 1/2\pi R_f f_{zf}$. Both C_c and C_i then relate to R_f as capacitances that define response break frequencies with that resistance. Substitution of the artificial C_c in the last C_f result produces the simplified

$$C_f = \sqrt{C_i C_c} \quad \text{for } \phi_m = 45° \text{ when } C_i \gg C_f$$

where $C_i = C_D + C_{id} + C_{icm}$, $C_c = 1/2\pi R_f f_c$, and f_c is the unity-gain crossover frequency of the op amp. Thus, the phase compensation capacitor C_f equals the geometric mean of the capacitance of the input circuit and the capacitance representing f_c. This result parallels the $f_i = \sqrt{f_{zf} f_c}$ expression in which f_i expresses the geometric mean of the analogous frequencies.

A second, more exact solution defines C_f for any value of photodiode capacitance but primarily serves cases where the $C_i \gg C_f$ approximation fails. Here, a simultaneous solution of two equations removes the interaction but produces a more complex equation for C_f. Solving the equations $C_f = 1/2\pi R_f\sqrt{f_{zf}f_c}$ and $f_{zf} = 1/2\pi R_f(C_i + C_f)$ for C_f yields

$$C_f = (C_c/2)\left(1 + \sqrt{1 + \frac{4C_i}{C_c}}\right) \quad \text{for } \phi_m = 45°$$

where $C_c = 1/2\pi R_f f_c$ and $C_i = C_D + C_{id} + C_{icm}$. Applications requiring this more exact C_f equation experience greater sensitivity to parasitic capacitances. There, parasitic effects remain significant compared to the smaller values of C_i and C_f. These parasitics alter the optimum value of C_f in either direction depending on the parasitic conditions of a given application. Some board parasitics add to the C_i term but others supplement C_f. Then, empirical adjustment resolves these unknowns with a final tuning of the phase compensation.

3.4 The Bandwidth Advantage of the Current-to-Voltage Converter

The phase compensation selected above defines the signal bandwidth for most photodiode amplifiers. Section 3.1.1 defines the exception to this rule for higher-gain amplifiers. For the general case, the current-to-voltage converter first extends bandwidth through the signal isolation described before. Then, two other factors improve bandwidth further through the circuit's residual gain peaking and an improved gain-bandwidth relationship.

As described above, the circuit's two-pole roll off produces gain peaking, and phase compensation reduces the peaking to preserve stability. Careful choice of this compensation permits bandwidth extension through the residual, compensated peaking. Figure 3.12 illustrates with an I-to-V response drawn to portray the $\phi_m = 45°$ compensation case. There, the gain peaking surrounding f_i pushes the response roll off to a slightly higher frequency than would occur with a single-pole roll off. This extends bandwidth beyond the circuit's f_i pole frequency to the extent permitted by acceptable gain peaking. For the traditional ± 3-dB bandwidth limit, examination of damping factors and corresponding peaking responses[3] shows a factor of 1.4 bandwidth increase for the 45° phase margin prescribed above. This margin restrains gain peaking to $+3$ dB followed by the final bandwidth limit at the -3-dB point of the response roll off. Thus, for the current-to-voltage converter with C_f set to break with R_f at f_i

$$BW = 1.4f_i = 1.4\sqrt{f_{zf}f_c}$$

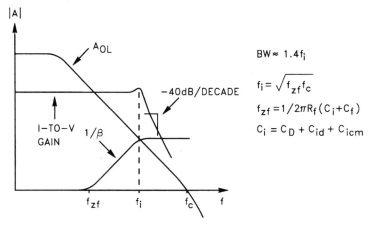

Figure 3.12 The two-pole response of the I-to-V gain produces peaking that extends bandwidth beyond f_i, and selecting phase compensation for $\phi_m = 45°$ makes BW $= 1.4 f_i$.

where $f_{zf} = 1/2\pi R_f(C_i + C_f)$, $C_i = C_D + C_{id} + C_{icm}$, and f_c is the unity-gain crossover frequency of the op amp.

This current-to-voltage converter bandwidth typically represents a dramatic improvement over the voltage-mode photodiode circuit described in Sec. 2.3. There, a single pole produces an overdamped response, fixing the -3-dB bandwidth at the pole frequency. Further, this pole frequency is much reduced from that produced by the current-to-voltage converter. There, input circuit capacitances break directly with the circuit load resistance, producing a bandwidth approximately equal to the f_{zf} frequency of the above equation. The current-to-voltage converter largely isolates the load resistor from those capacitances, extending bandwidth to $1.4\sqrt{f_{zf}f_c}$. Comparing the two bandwidth limits shows a factor of $1.4\sqrt{f_c/f_{zf}}$ improvement for the current-to-voltage converter. Thus, choosing an op amp with $f_c \gg f_{zf}$ assures a significant bandwidth improvement.

The bandwidth of the current-to-voltage converter also benefits from a more favorable gain-bandwidth relationship that reduces the bandwidth's sensitivity to gain increases. Normally, the constant gain-bandwidth product of an op amp produces a bandwidth reduction directly proportional to any gain increase. Also, the previous voltage-mode circuit produces this same gain-bandwidth relationship. However, for photodiode applications, the current-to-voltage converter replaces this direct relationship with a square root function. Increasing the R_f feedback resistance of the circuit still increases the circuit's gain, $e_o/i_p = R_f$, in direct proportion. The corresponding bandwidth decreases as reflected through BW $= 1.4\sqrt{f_{zf}f_c}$ where

$f_{zf} = 1/2\pi R_f(C_i + C_f)$. Here, f_{zf} decreases in a direct proportion to the increased R_f. However, the bandwidth only decreases in proportion to $\sqrt{f_{zf}}$ or $\sqrt{R_f}$. For the current-to-voltage converter, gain-bandwidth product is simply GBW = R_f(BW), and substituting BW from its equation above shows this product to be GBW = $1.4\sqrt{R_f f_c/2\pi(C_i + C_f)}$. Thus, the maximum gain-bandwidth product results from the maximum practical value of R_f. Above a certain R_f value, parasitic capacitance rolls off the gain provided by this resistor in the independent bandwidth limit described in Sec. 3.1.1.

The response plots of Fig. 3.11 provide visual insight into the improved gain-bandwidth relationship. There, the intercept of the $1/\beta$ and A_{OL} responses defines f_i, the pole frequency fundamental to the BW = $1.4f_i$ response limit. As previously discussed, two curves represent the $1/\beta$ response with one curve being solid and the other dashed. Consider starting with the dashed curve and moving back to the solid one. To do this, increasing R_f moves f_{zf} back in frequency with direct proportionality. This also shifts the $1/\beta$ curve and its f_i intercept frequency backward but not in direct proportion. The intercept recedes more gradually because the A_{OL} curve rises as f_i decreases, reducing the decrease in the f_i bandwidth limit.

This favorable gain-bandwidth characteristic also permits optimization through increased photodiode area. Increasing diode area increases the overall circuit response to a light source at a rate greater than the accompanying decline in bandwidth. The photodiode output and the diode capacitance both increase in direct proportion to the diode area A_D. Increasing A_D produces a directly proportional increase in the light-to-voltage gain of the circuit. However, the bandwidth described by the previous equations declines only by the square-root of C_D. This makes the gain-bandwidth product for the overall circuit proportional to $\sqrt{A_D}$. The maximum light-to-voltage gain-bandwidth product results in a photodiode area that spans as much of the area illuminated by the light source as is practical.

To maximize bandwidth instead of gain-bandwidth product, choose R_f to utilize the full amplifier bandwidth. Making R_f smaller moves f_{zf} and f_i to the right in Fig. 3.11 up to the limit imposed by f_c. Beyond f_c, the amplifier lacks the bandwidth required for further extension of the current-to-voltage converter response. Then, for maximum bandwidth, selecting R_f places the intercept frequency f_i at the amplifier unity crossover f_c. This moves the $1/\beta$ curve fully to the right, compressing its response rise to zero. Then, the three frequencies defining the circuit response coincide at $f_{zf} = f_i = f_c$. Given this condition, the expression for f_{zf} sets the circuit's feedback resistor to

$$R_f = 1/2\pi f_c C_i \quad \text{for maximum bandwidth}$$

where $C_i = C_D + C_{id} + C_{icm}$. Note that this expression does not contain C_f because the removal of the $1/\beta$ rise also removes the need for phase compensation. Any further increase in bandwidth must come from the use of a higher-speed op amp that moves the f_c limit to a higher frequency. Once again, a square root relationship governs this final improvement, making the bandwidth increase proportional to $\sqrt{f_c}$.

3.5 Phase Compensation Alternatives

The earlier phase compensation discussion of this chapter produces a generally accepted compromise between bandwidth and signal gain peaking. Here, the phrase "signal gain peaking" differentiates the peaking that affects the signal response from the noise gain peaking described in Chap. 5. Design equations developed earlier in this chapter place the C_f compensation pole at the $1/\beta$ intercept to first ensure 45° of phase margin for circuit stability. This phase margin also limits signal gain peaking to 3 dB prior to the signal response roll off, optimizing the ±3-dB bandwidth. Some applications require the reduced peaking provided by a more conservative C_f selection. This section develops a C_f design equation directly linked to a user-specified value of signal gain peaking. Then, specifying a peaking value first defines the corresponding damping factor of the circuit response. Then, this factor defines the C_f capacitance required to produce the specified peaking.

The more precise peaking control here often encounters practical limitations imposed by parasitic capacitances. Precise control may require very small capacitance values or capacitance tuning to optimize the bandwidth-versus-peaking compromise. For small capacitance values, the parasitic variable initially suboptimizes the bandwidth result by requiring the guard-banded selection of a larger C_f value. Capacitance tuning removes this variable, but just the parasitic capacitance introduced by the tuning operation still complicates the small C_f case. Tee networks remove these limitations by adapting the photodiode amplifier to larger C_f levels, reducing the sensitivity to parasitic capacitances.

3.5.1 The photodiode amplifier's second-order response

To develop the more precise C_f equation, the analysis below relies upon the well-known peaking characteristics of a second-order or two-pole system. There, the damping ratio ζ serves as a figure of merit in signal gain peaking evaluation. Figure 3.13 illustrates the peaking-versus-ζ compromise in an overlay of frequency response curves corresponding to different ζ values. There, increasing ζ reduces signal gain peaking but also reduces bandwidth. Making $\zeta = 1$ produces maximum

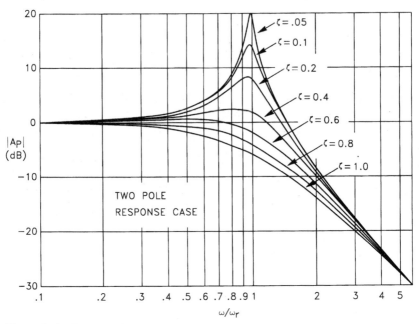

Figure 3.13 Damping factor ζ serves as a figure of merit reflecting response gain peaking.

damping, and no peaking, but excessively limits bandwidth. This case represents the single-pole roll off case automatically produced by a first-order system. A second-order system, like the current-to-voltage converter, offers greater bandwidth under less conservative damping conditions. Also, examination of the figure's response curves reveals that reducing ζ increases bandwidth without introducing peaking up to a point. Later analysis defines the optimum as $\zeta = 0.707$, a familiar fraction, for maximum bandwidth from a second-order system without signal gain peaking. This damping ratio produces a critically damped response. The 3-dB peaking response of the previous phase compensation discussions results for $\zeta = 0.384$.

Comparing the response equations of the current-to-voltage converter and the standard second-order system first translates well-known response characteristics to the current-to-voltage converter case. This comparison identifies the converter's damping ratio and natural frequency in terms of the converter's circuit components. Then, further analysis defines the relationship between the damping ratio and the magnitude of signal gain peaking. Finally, a user-specified gain-peaking limit defines a damping ratio that in turn defines a relationship between the converter's circuit components. This latter relationship yields the value of phase compensation capacitor C_f required to limit gain peaking to that specified.

Starting the process, analysis of Fig. 3.14 first defines the current-to-voltage converter's signal response in terms of the circuit elements. This circuit follows from the model originally presented in Fig. 3.7. As before, $C_i = C_D + C_{id} + C_{icm}$ represents the combined capacitances of the photodiode and the differential and common-mode input capacitances of the op amp. Ideally, the model shown produces an output signal equal to the photodiode current times the feedback impedance $i_p Z_f$, where Z_f equals the impedance combination of R_f and C_f. Amplifier gain error subtracts from this ideal output through the amplified effect of error signal e_o/A_{OL}. The finite open-loop gain A_{OL} of the op amp requires this error signal between the op amp inputs to support the output signal e_o. The circuit amplifies e_o/A_{OL} by the gain set by the impedances of the feedback and input circuits, $1 + Z_f/Z_i$, where Z_i equals the impedance of C_i. Writing a loop equation for output signal e_o produces

$$e_o = i_p Z_f - \left(1 + \frac{Z_f}{Z_i}\right)\left(\frac{e_o}{A_{OL}}\right)$$

where $Z_f = R_f/(1 + R_f C_f s)$, $Z_i = 1/C_i s$, and $C_i = C_D + C_{id} + C_{icm}$.

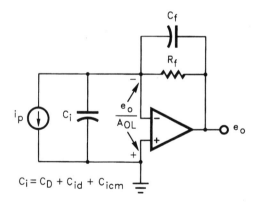

$$C_f = \left(2\zeta^2 - 1 + 2\zeta\sqrt{\zeta^2 - 1 + \frac{C_i}{C_c}}\right) C_c$$

where $\zeta = \left(\sqrt{1 + 1/A_{Pn}} - \sqrt{1 - 1/A_{Pn}}\right)/2$,

$A_{Pn} = |A_{CL}|_P/A_{CLi}$ and $C_c = 1/2\pi R_f f_c$

Figure 3.14 For the current-to-voltage converter, a specified gain peaking ratio A_{Pn} defines the corresponding damping ratio ζ and the required phase compensation C_f.

Over most of its useful frequency range, the open-loop gain A_{OL} above follows a single-pole response approximated by $A_{OL} = \omega_c/s$, where ω_c is the unity-gain crossover frequency of the op amp. This gain approximation produces good experimental agreement for photodiode amplifiers because the amplifier's $1/\beta$ intercept virtually always occurs in this single-pole region. Substituting the above expressions for Z_f, Z_i, and A_{OL} in the previous e_o equation and solving for e_o produces the final s plane response:

$$e_o = \frac{i_p R_f}{s^2/\omega_{zf}\omega_c + s\left(1/\omega_{pf} + 1/\omega_c\right) + 1}$$

where $\omega_{zf} = 1/R_f(C_i + C_f)$ and $\omega_{pf} = 1/R_f C_f$.

Examination of the above equations reveals several anticipatable characteristics. First, ω_{zf} and ω_{pf} represent the feedback zero and pole that previously shaped the circuit's $1/\beta$ response. As described before, the circuit amplifies the e_o/A_{OL} error signal by a gain equal to $1/\beta$ up to the roll off limit imposed by the amplifier's open-loop response. This limit reflects in the presence of ω_c in the e_o equation. Also, the s^2 denominator term of the equation shows that the photodiode amplifier produces a two-pole or second-order response, permitting direct comparison with the standard second-order system results.

3.5.2 Translating the second-order results to the photodiode amplifier

Reducing the photodiode amplifier's response expression to the standard form of a second-order system transfers known results to this case. Then, comparison of response expression terms defines the photodiode amplifier's fundamental response characteristics in terms of those results. For this comparison, substituting $s = j\omega$ in the above e_o equation, and solving for $A_{CL} = e_o/i_p$ reduces the photodiode amplifier's response to the standard form of the two-pole system

$$A_{CL} = \frac{A_{CLi}}{1 - \omega^2/\omega_{zf}\omega_c + j\omega\left(1/\omega_{pf} + 1/\omega_c\right)}$$

where $A_{CLi} = R_f$ is the ideal closed-loop gain.

For the generalized two-pole system, standardized results define closed-loop gain as[4]

$$A_{CL} = \frac{A_{CLi}}{1 - \omega^2/\omega_r^2 + j2\zeta\omega/\omega_r}$$

where ω_r is the natural resonant frequency, and ζ is the damping ratio of the A_{CL} response. Comparison of the two A_{CL} responses above then defines ω_r and ζ for the photodiode amplifier. First, $\omega_r = \sqrt{\omega_{zf}\omega_c}$

Bandwidth and Stability   57

or $f_r = \sqrt{f_{zf}f_c}$ repeats a familiar frequency for the photodiode amplifier. From before, the amplifier's $1/\beta$ curve intercepts the amplifier's A_{OL} response at this same frequency $f_i = \sqrt{f_{zf}f_c}$. Thus, the circuit's natural resonant frequency equals the frequency of the $1/\beta$ intercept, as expected from earlier resonance discussions. As described there, photodiode amplifiers frequently oscillate at $f_i = f_r$ unless phase compensation degenerates the natural resonance of the two-pole system.

A second comparison of the two A_{CL} expressions above defines the damping ratio to quantify the signal gain peaking. Equating the j terms of the A_{CL} denominators above and solving for ζ yields

$$\zeta = \frac{1 + \omega_c/\omega_{pf}}{2\sqrt{\omega_c/\omega_{zf}}} = \frac{1 + f_c/f_{pf}}{2\sqrt{f_c/f_{zf}}}$$

where $f_{pf} = 1/2\pi R_f C_f$, $f_{zf} = 1/2\pi R_f (C_i + C_f)$, and $C_i = C_D + C_{id} + C_{icm}$ from before. From these equations, note that increasing phase compensation capacitance C_f reduces f_{pf} to increase the damping ratio. Also, increasing the photodiode capacitance C_D increases C_i, reducing f_{zf} and decreasing the damping. Thus, signal gain peaking decreases with an increased C_f and increases with an increased C_D, as expected.

To quantify these effects, further analysis relates the damping ratio ζ to the gain peaking. For the second-order system, solving for the magnitude of its A_{CL} gain produces

$$|A_{CL}| = \frac{A_{CLi}}{\sqrt{1 - \omega^2/\omega_r^2 + 4\zeta^2\omega^2/\omega_r^2}}$$

Differentiating this expression with respect to ω and equating the result to zero defines the frequency of the $|A_{CL}|$ peak magnitude as $\omega_P = \omega_r\sqrt{1 - 2\zeta^2}$. Then, substituting this ω_P expression for ω in the above $|A_{CL}|$ expression defines the peak gain magnitude as

$$|A_{CL}|_P = \frac{A_{CLi}}{2\zeta\sqrt{1 - 2\zeta^2}}$$

This result expresses the peak gain magnitude solely as a function of the ideal gain A_{CLi} and the damping ratio ζ.

3.5.3 Selecting the phase compensation for limited peaking

For design purposes, a more useful result expresses the ζ ratio required to limit gain peaking to a specific but relative level. Solving the $|A_{CL}|_P$ expression for ζ defines the damping ratio as a function of a specific $|A_{CL}|_P$. Then, dividing $|A_{CL}|_P$ by A_{CLi} normalizes the result to a relative level, removing the A_{CLi} variable, and $A_{\text{Pn}} = |A_{\text{CL}}|_P / A_{\text{CLi}}$. This analysis process produces the damping ratio required to deliver a specified, fractional gain peak independent of the ideal closed-loop gain. That damping ratio is

$$\zeta = \frac{\sqrt{1 + 1/A_{\text{Pn}}} - \sqrt{1 - 1/A_{\text{Pn}}}}{2}$$

where $A_{\text{Pn}} = |A_{\text{CL}}|_p / A_{\text{CLi}}$ is the normalized gain peak. Note that for purposes of A_{Pn} here, $|A_{\text{CL}}|_p$ and A_{CLi} represent linear rather than logarithmic expressions of gain. For example, a 3-dB or 41 percent peak makes $A_{\text{Pn}} = 1.41$ and $\zeta = 0.384$.

Given the expression for ζ above, the phase compensation required to produce a given damping ratio defines C_f for the reduced-peaking current-to-voltage converter. Further analysis defines the C_f design equation beginning with the previous result

$$\zeta = \frac{1 + f_c/f_{\text{pf}}}{2\sqrt{f_c/f_{\text{zf}}}}$$

where $f_{\text{pf}} = 1/2\pi R_f C_f$ and $f_{\text{zf}} = 1/2\pi R_f (C_i + C_f)$. Substituting the expressions for f_{pf} and f_{zf} and solving for C_f yields the final design equation with a simplification produced by defining a fictitious C_c capacitance. As described in Sec. 3.3.5, defining $C_c = 1/2\pi R_f f_c$ indicates that value of capacitance that would result from a break frequency with R_f at the amplifier's unity crossover frequency f_c. This break parallels those produced by other circuit capacitances with R_f. Then, solving for C_f yields

$$C_f = \left(2\zeta^2 - 1 + 2\zeta\sqrt{\zeta^2 - 1 + C_i/C_c}\right) C_c$$

where $\zeta = \left[\sqrt{1 + 1/A_{\text{Pn}}} - \sqrt{1 - 1/A_{\text{Pn}}}\right]/2$, $A_{\text{Pn}} = |A_{\text{CL}}|_p / A_{\text{CLi}}$, $C_i = C_D + C_{\text{id}} + C_{\text{icm}}$, and $C_c = 1/2\pi R_f f_c$. To apply this phase compensation equation, the desired value of normalized gain peaking first defines A_{Pn}. Then, A_{Pn} defines ζ for substitution in the above C_f equation.

As expected, more conservative damping also reduces bandwidth. In Sec. 3.3, the 45° phase margin guideline made two bandwidth limits coincident so that neither dominated the final result. The two limits result from the $1/\beta$ intercept at f_i and the pole produced by C_f at f_{Cf}. Increasing C_f to increase damping moves the f_{Cf} pole to a lower frequency, making f_{Cf} the dominant bandwidth limit. Typically, the f_i limit still influences bandwidth through residual gain peaking, but neglecting this influence permits a simple first-order approximation to the result. Then, for $\phi_m > 45°$ or $\zeta > 0.384$, BW $\approx f_{\text{Cf}} = 1/2\pi R_f C_f$ approximates the −3-dB bandwidth.

3.5.4 Realizing the precise phase compensation

Precise control of gain peaking sometimes requires a similar control over small values of C_f. There, large values of feedback resistance R_f reduce the optimum value of C_f to such low levels that parasitic capacitance variables significantly alter the phase compensation result. In other cases, the tolerance variations of circuit capacitances make C_f tuning desirable for precise peaking control. However, the tuning process introduces parasitic mutual capacitances that further complicate the low-level C_f case. Tee networks, both resistive and capacitive, resolve these problems by increasing the compensation capacitance values to levels that are less sensitive to parasitic effects. Also, the capacitive tee facilitates a shielding ground connection that removes mutual capacitance effects in the tuning operation. While providing these conveniences, tee networks also increase noise gain, but judicious design choices limit the increase.

First, consider the effects of converting a large feedback resistance to a resistive tee network formed of lower-valued resistances. Section 2.2.2 describes this alternative where the tee network reduces offset error. For the more general photodiode amplifier, the resistor tee also permits larger capacitance values for C_f as illustrated by Fig. 3.15. There, the R_1/R_2 voltage divider attenuates the

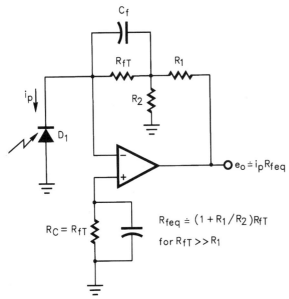

Figure 3.15 Replacing the feedback resistor with a tee reduces resistance levels, permitting phase compensation with a larger capacitor having less sensitivity to parasitic capacitances.

feedback signal to effectively multiply the basic R_{fT} feedback resistance. For $R_{fT} \gg R_1$, this produces an equivalent feedback resistance of $R_{feq} \approx (1 + R_1/R_2)R_{fT}$. Then, $e_o = i_p R_{feq}$ and the smaller R_{fT} produces the same current-to-voltage gain as a larger R_f does in the basic circuit. The associated phase compensation bypass of the smaller R_{fT} resistance requires a larger and less parasitic-sensitive value of C_f for the same compensation result. As seen from $R_{feq} \approx (1 + R_1/R_2)R_{fT}$, the tee multiplies the effective value of R_{fT} by a factor of $1 + R_1/R_2$. For equivalent phase compensation results C_f increases by this same factor. Thus, all previous phase compensation results transfer to this tee case with the optimum value of C_f simply increased by the factor $1 + R_1/R_2$.

Other aspects of the resistive feedback tee potentially alter the current-to-voltage converter performance. Just the structure of the resistor tee provides a first improvement to the feedback bypass task. The added physical spacing of the tee, with three resistors instead of one, reduces the net stray capacitance across the feedback path. Also, the stray capacitance across each resistor produces much less effect due to the lower resistance values. However, the circuit's sensitivity to other stray capacitance from the op amp output to its input remains the same as with a single resistor feedback. In addition, C_f only bypasses R_{fT} here, leaving the parallel combination of R_1 and R_2 to react with the circuit's input capacitance at higher frequencies. However, making R_1 and R_2 small resistances easily moves this reaction beyond the response range of the op amp.

Alternately, a capacitive feedback tee network bypasses a single resistor R_f with even subpicofarad equivalent capacitances and with a shielding immunity to the parasitic capacitance of the tuning operation. Shown in Fig. 3.16, this tee bypasses R_f with capacitor values much greater than that of a single capacitor bypass. To do so, the tee uses a capacitive divider, formed with C_2 and C_3, to attenuate the bypass signal applied to C_1 at the circuit input. With only a fraction of the output signal impressed upon C_1, this capacitor supplies far less shunting current to the input node, as would a much smaller feedback capacitor. In effect, the capacitor tee produces an equivalent feedback capacitance of

$$C_{feq} = \frac{C_1 C_2}{C_1 + C_2 + C_3}$$

This equivalent capacitance shunts R_f with a response pole at $f_{Cf} = 1/2\pi R_f C_{feq}$.

The capacitor tee also reduces tuning sensitivity to stray capacitance in two ways. First, the tunable capacitor shown, C_3, can now be a

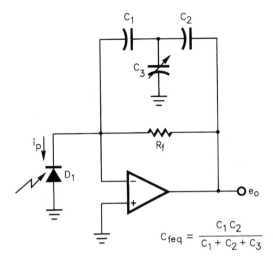

Figure 3.16 Replacing the feedback capacitor with a tee also reduces sensitivity to parasitic capacitances through the increased capacitance values permitted by the tee connection.

$$C_{feq} = \frac{C_1 C_2}{C_1 + C_2 + C_3}$$

larger capacitor, and thus, is less susceptible to the parasitic effects introduced by the tuning operation. Also, this capacitor returns to ground, offering a shielding opportunity to further reduce the tuning variability.

References

1. J. Graeme, "Feedback Plots Offer Insight into Operational Amplifiers," *EDN*, January 19, 1989, p. 131.
2. J. Graeme, "Feedback Models Reduce Op Amp Circuits to Voltage Dividers," *EDN*, June 20, 1991, p. 139.
3. G. Tobey, J. Graeme and L. Huelsman, *Operational Amplifiers: Design and Applications*, McGraw-Hill, New York, 1971.
4. M. Van Valkenburg, *Network Analysis*, Prentice-Hall, Englewood Cliffs, N.J., 1965.

Chapter

4

Wideband Photodiode Amplifiers

Junction capacitance severely restricts the bandwidth of the basic photodiode as described in Sec. 2.3. There, any signal voltage developed by the diode reacts with this capacitance, shunting the diode's output current. Three op amp circuit methods greatly ease this restriction through signal isolation, photodiode bias, and photodiode bootstrap.[1,2] Signal isolation removes the signal voltage from the photodiode through the op amp current-to-voltage converter as described in Chaps. 2 and 3. This circuit holds the diode voltage near zero while supplying the signal voltage at the op amp output. Section 3.4 quantifies the resulting bandwidth improvement.

For even greater improvement, biasing or bootstrapping the photodiode further reduces the effect of the diode's capacitance. A reverse bias voltage applied to the photodiode improves bandwidth by reducing the diode junction's capacitance. Bootstrap of the photodiode improves bandwidth in much the same way as the current-to-voltage converter by isolating the diode capacitance from the signal voltage. However, for small photodiodes, the bootstrap alternative achieves greater bandwidth. Combining bootstrap with the current-to-voltage converter makes bandwidth almost immune to the capacitance of the photodiode. However, this combination requires a bootstrap amplifier that meets several stringent requirements.

4.1 Biasing the Photodiode

Even with the current-to-voltage converter, photodiode capacitance remains a primary limitation to bandwidth as described in Sec. 3.4. Simple reverse bias of the photodiode reduces this capacitance through the capacitance's voltage dependence. However, this simple bias also produces dominant offset and noise errors. Voltage across the diode induces a leakage current that typically far exceeds that of the op amp input. Similarly, the noise of the bias source typically far exceeds the amplifier's input noise voltage. These added leakage and noise sources produce offset and noise at the circuit output. Simple R-C filtering of the bias voltage removes much of the added noise. However, both the resistance and the capacitance of this filter must be selected with care to make this filtering effective. Alternately, converting to a differential-input circuit configuration lets the common-mode rejection (CMR) of the op amp attenuate both the offset and noise errors. There, a dummy photodiode couples canceling leakage and noise effects to the amplifier.

4.1.1 The bias effect

Reverse bias actually increases the speed of a photoreceiver in two ways. In addition to reducing diode capacitance, the bias increases the response speed of the photodiode as described in Chap. 1. The bias expands the width of the diode's internal depletion region to produce both effects. First, the expanded depletion regions encompasses more of the diode's photo-generated carriers in this region of electric field gradient. There, the field gradient accelerates the carriers, moving them to the diode terminals more rapidly. This acceleration increases the diode response speed, but the bandwidth of the overall photodetector circuit typically remains limited by the response of the photodiode amplifier. There, the capacitance of the photodiode reduces the amplifier bandwidth as described in Chap. 3.

Under bias, the diode's expanded depletion region also reduces this capacitance. The increased width of the region corresponds directly to an increased thickness of the junction capacitance's effective dielectric layer. Increased dielectric thickness corresponds to decreased capacitance, and this junction capacitance responds to a reverse-bias voltage V_R as expressed by

$$C_D = \frac{C_{D0}}{\sqrt{1 + V_R/\varphi_B}}$$

Here, C_{D0} is the photodiode capacitance at zero bias, and φ_B is the built-in voltage of the diode junction. Applying a reverse bias V_R reduces C_D from its C_{D0} maximum through a comparative relationship with φ_B. For silicon photodiodes, $\varphi_B \approx 0.6$ V and making $V_R \gg \varphi_B$ produces the desired capacitance reduction. For example, making $V_R = 10$ V reduces the capacitance by a factor of 4.2 from its C_{D0} value.

From Sec. 3.4, the basic photodiode amplifier's bandwidth responds to this capacitance reduction as expressed through

$$\text{BW} = 1.4 f_i = 1.4\sqrt{f_{zf}f_c}$$

where $f_{zf} = 1/2\pi R_f(C_i + C_f)$, $C_i = C_D + C_{id} + C_{icm}$, and f_c is the unity-gain crossover frequency of the op amp. For large photodiode capacitances, $C_i \approx C_D$ and $f_{zf} \approx 1/2\pi R_f C_D$. Then, BW $\propto 1/\sqrt{C_D}$ and the 4.2:1 reduction in C_D produced by $V_R = 10$ V improves bandwidth by about a factor of 2.

4.1.2 Photodiode biasing with the current-to-voltage converter

With this circuit, simply returning the diode to a voltage source instead of common introduces the desired reverse bias. This modified return also introduces new error sources, but later circuit alternatives greatly reduce these. Figure 4.1 illustrates the basic photodiode bias with source V_B. There, the op amp input holds the anode of D_1 at zero voltage, and dc voltage source V_B sets the reverse bias at $V_R = V_B$. This reduces the diode capacitance, as described above but increases dc offset and noise errors. Voltage source V_B induces a diode leakage current and introduces a noise current in the diode capacitance. Figure 4.2 models these effects with a diode leakage current I_L and a capacitive noise current i_{nB}. Flow of these error currents in feedback resistor R_f produces output errors that often dominate offset and noise performance.

For the biased photodiode case, the I_L leakage current typically overwhelms dc offset error. To significantly reduce photodiode capacitance, the diode reverse bias must be large, and this raises diode leakage to its full saturation level. With typical photodiodes, this leakage current far exceeds the FET leakage that produces the amplifier input current. In the absence of diode bias $V_B = 0$, the photodiode resides

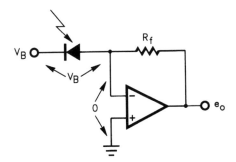

Figure 4.1 The current-to-voltage converter's virtual-ground input permits simple control of photodiode bias for diode capacitance reduction.

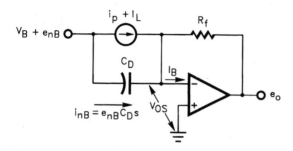

$$e_o = -(i_p + I_L - I_B + e_{nB}C_D s)R_f + V_{OS}$$

Figure 4.2 The photodiode bias of Fig. 4.1 introduces offset and noise errors through diode leakage current I_L and bias-source noise voltage e_{nB}.

across the op amp inputs with virtually no voltage to support a diode leakage current. Then, the amplifier input current I_B dominates the circuit's input error current and produces an output offset voltage of $I_B R_f$. Choosing a FET-input op amp minimizes this error, but it remains a major offset component for high-gain photodiode amplifiers. Adding to this component, the input offset voltage of the op amp V_{OS} produces a similar output offset.

However, with a typical biased photodiode and a FET-input op amp, $I_L \gg I_B$. These amplifiers produce an I_B input current that also results from junction leakage. The I_L and I_B currents differ in magnitude in accordance with the associated junction voltages and the relative junction areas. A large physical area enhances the photodiode sensitivity, favoring this bias in the photodiode selection. However, leakage current increases in proportion to this area, making the circuit offset highly sensitive to any diode bias. The FETs of the amplifier inherently support a bias voltage but with junction areas as much as 1000 times smaller. There, small junction areas intentionally limit amplifier input leakage and input capacitance. Thus, even a moderate-size photodiode produces a dominant leakage current in the presence of diode bias. Often, the resulting offset error increases by a factor far greater than that of the bandwidth improvement achieved.

Noise also increases with photodiode bias through two added noise sources. First, the noise component of the diode leakage current $\sqrt{2qI_D}$ significantly increases the output noise component resulting from input noise current. Chapter 5 describes this component along with the others that contribute to the net output noise signal. More significantly, the V_B bias source impresses its noise voltage component e_{nB} across the diode capacitance. In Fig. 4.2, e_{nB} produces a noise current $e_{nB}C_D s$ in capacitance C_D. This current flows through R_f, creating an output component of noise $e_{nB}R_f C_D s$. Thus, e_{nB} experiences a noise gain of $A_{nB} = R_f C_D s$ with a response zero at $1/2\pi R_f C_D$. This zero produces a high level of higher-frequency gain to greatly amplify the e_{nB} signal. Chapter 5 also describes the associated gain peaking that often makes the amplifier's input noise voltage a dominant effect. Practical bias sources produce e_{nB} noise levels that easily surpass this amplifier noise, making the gain peaking effect even more serious. Then, biasing the photodiode will typically increase output noise by as much as a factor of 100 unless other measures reduce this noise effect.

4.2 Bias Improvements

Two modifications to photodiode bias reduce its error effects through filtering and CMR. Filtering removes the noise effect, and CMR removes both the noise and offset. However, activating the op amp's CMR requires the addition of a second photodiode.

4.2.1 Filtering the bias voltage

Fortunately, circuit conditions make this filtering task relatively simple, but the filter component selection requires some attention. Figure 4.3 shows the filtered version of the voltage-bias case with a low-pass filter inserted between bias source V_B and the photodiode. While simple in structure, this filter produces subtle effects that direct its design through the two guidelines $R_B \ll R_f$ and $C_B \gg C_D$. Then, the filter supplies a dc bias voltage $V_B' \approx V_B$ with little voltage loss

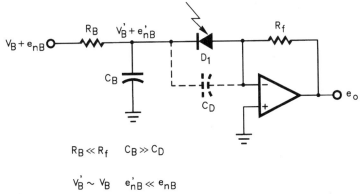

Figure 4.3 Simple R-C filtering greatly attenuates the noise effect of the V_B bias source when the filter's resistance and capacitance impedances remain small compared to those of the circuit.

while significantly attenuating the ac noise signal to $e_{nB}' \ll e_{nB}$. The selection of R_B assures the bias transfer, and the selection of C_B assures the noise reduction.

Making $R_B \ll R_f$ ensures that $V_B' \approx V_B$ in spite of the transfer error introduced by the diode's current. This current flows in the filter, producing bias and signal errors in V_B'. The leakage component of the diode current produces a dc voltage drop on R_B but typically reduces the transferred bias voltage by only a small amount. However, the signal component of the diode current produces an ac modulation of V_B', reintroducing some of the nonlinearity originally removed by the current-to-voltage converter. As described in Chap. 2, this converter removes load voltage swing from the diode, ensuring a stable diode voltage and a corresponding constant responsivity to light. Adding the filter to this converter reintroduces signal swing across the diode due to the reaction of the diode's current with the filter's output impedance. Any signal voltage developed with this impedance appears directly across the photodiode since the other end of the diode returns to the virtual ground of the amplifier input. At low frequencies, the filter output impedance reaches its maximum, at $R_o = R_B$, and produces its greatest modulation of V_B'. Filter resistor R_B conducts the same signal current as R_f and reproduces a portion of the output signal voltage across the diode. Making $R_B \ll R_f$ limits the modulation voltage to a small fraction of the load voltage developed upon R_f.

Similarly, making $C_B \gg C_D$ avoids the otherwise overriding noise effect of the bias source. Both a circuit evaluation and a noise gain analysis support this C_B choice. In the circuit, the two capacitors shown effectively appear in parallel, due to the virtual ground return of C_D at the amplifier's inverting input. Thus, the attenuated noise signal of the bias source, e_{nB}', drives the two capacitances equally. Resulting capacitive currents produce the voltage drop on R_B that reduces the original e_{nB} to the attenuated e_{nB}'. Increasing the value of C_B increases this current, and the resulting noise attenuation to reduce the residual e_{nB}' left to drive C_D. In effect, C_B and C_D form a current divider, sharing the task of conducting noise current through R_B to attenuate e_{nB}'. Capacitance C_B diverts its current to ground, removing the associated noise effect. Capacitance C_D, however, delivers its current to R_f, producing an output noise voltage. Making $C_B \gg C_D$ diverts the majority of the noise current to ground, minimizing the noise effect of the original e_{nB} noise source.

Noise gain analysis reinforces this conclusion and combines it with the earlier condition of $R_B \ll R_f$. From the Fig. 4.2 discussion, the unfiltered case presented a noise gain of $A_{nB} = R_f C_D s$ to e_{nB}. Analyzing Fig. 4.3's transfer response to noise signal e_{nB} produces the modified noise gain

$$A_{\mathrm{nB}}' = \frac{e_{\mathrm{onB}}}{e_{\mathrm{nB}}} = \frac{R_f C_d s}{1 + R_B(C_B + C_D)s}$$

where e_{onB} is the output noise resulting from the noise of the bias source e_{nB}. As desired, the presence of the R_B and C_B filter introduces a response pole to this gain, counteracting the inherent zero produced by R_f and C_D. That zero previously increased the high-frequency noise gain, potentially dominating noise performance. At higher frequencies, with $C_B \gg C_D$, the filtered noise gain here becomes

$$A_{\mathrm{nBH}}' = \frac{R_f C_D}{R_B C_B}$$

The division ratio of this expression shows that the previous requirement of $R_B \ll R_f$ tends to make this noise gain large. However, the circuit requires this resistance condition to avoid signal modulation of the diode voltage. Counteracting this large gain tendency, making $C_B \gg C_D$ restores the filter's effectiveness in reducing the noise response. Even greater noise reduction results from a combination of this filtering with the CMR described below.

4.2.2 Common-mode rejection of the bias-induced errors

As described earlier, photodiode bias reduces diode capacitance, increasing bandwidth, but introduces new noise and offset errors. The low-pass filtering described above attenuates, but does not remove, the new noise error. Further, the filtering described fails to address the dc offset error introduced by the increased leakage current of the photodiode. Making use of the op amp's differential inputs greatly reduces both new errors through the amplifier's CMR. To activate this rejection, Fig. 4.4 adds a second matching photodiode and a second matching resistor. There, only the original photodiode D_1 remains open to light input, and blocking D_2 from the light source provides error cancellation without also producing signal cancellation.

Then, only D_1 supplies a signal current, but both diodes supply leakage and noise currents to the op amp. Currents supplied by the two diodes drive opposite-polarity amplifier inputs, and circuit matching makes the diode error currents produce counteracting effects. First, component matching assures equal diode error currents and equal current-to-voltage conversion results. Next, the common dc bias of the circuit's two diodes assures matching error responses to this bias. In the circuit, the anodes of two diodes connect to the op amp inputs

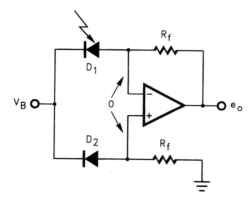

Figure 4.4 Adding a dummy diode and a second R_f resistor produces a differential-input connection for CMR of the bias-induced errors.

where feedback establishes equal anode potentials. Also, the cathodes connect together at the bias source, establishing equal cathode potentials. Thus, the two diodes support equal voltage drops whether from the dc or noise outputs of the bias source.

Figure 4.5 models the signal and error effects of these matching circuit conditions. There, only the current from D_1 includes the signal component i_p to produce an uncanceled output signal $-i_p R_f$. The other current components produce canceling output effects through equal current effects. The dc component of the bias voltage V_B produces matching leakage currents I_L in the two diodes. Similarly, the noise

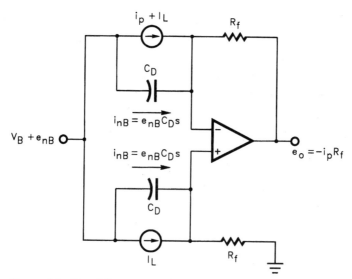

Figure 4.5 The differential-input connection of Fig. 4.4 results in matching I_L and i_{nB} error currents that produce canceling effects.

component of V_B, e_{nB}, produces matching noise currents of $e_{nB}C_D s$ in the diode capacitances. Flow of these matched currents in the two R_f resistors produces equal resistor voltages at opposite amplifier inputs. Simple analysis shows that these equal error voltages produce canceling effects in the circuit output voltage. With two photodiodes from the same manufacturing lot, this circuit's matching equalizes the leakage and noise errors to within about a 5 percent tolerance for a 20:1 error reduction. However, even with this circuit's CMR, photodiode bias still compromises errors when compared to the original zero-biased connection. Bootstrap alternatives frequently offer a better solution for improved bandwidth.

4.3 Bootstrapping the Photodiode

Bootstrap drive of a photodiode improves bandwidth over that achieved with the basic current-to-voltage converter. As described in Sec. 2.3, the basic current-to-voltage converter improves bandwidth by removing load signal voltage from the capacitance of the photodiode source. This avoids capacitive currents that otherwise absorb the diode's signal current at higher frequencies. Bootstrap drive also removes load signal swing from a signal source as commonly applied in voltage-mode circuits. Analogous bootstrap techniques address the current-to-voltage mode of photodiode amplifiers. Then, the bootstrap circuit either replaces or works with the basic current-to-voltage converter. In both cases, bandwidth improves without the need for photodiode bias. Replacing the current-to-voltage converter with a bootstrap amplifier improves bandwidth for the case of small diode capacitance and adding bootstrap to the converter addresses the larger capacitance case.

4.3.1 The basic bootstrap photodiode amplifier

Conventional bootstrap drives the common return of a voltage source with the voltage developed on the circuit load. This concept translates directly to photodiode sources with results very similar to those described for the current-to-voltage converter. Phase compensation requirements and the resulting bandwidth closely follow the earlier discussion but with added bandwidth in low-capacitance cases.

A comparison of voltage-mode and bootstrap photodiode connections reveals the bootstrap benefit. For comparison, Fig. 4.6 repeats the voltage-mode circuit originally described in Sec. 2.3. With this circuit, the photodiode directly drives a resistive load to produce the circuit's output signal voltage e_o. Unfortunately, that voltage also appears across the diode and its capacitance C_D, developing a parasitic capacitive current i_c. At higher frequencies, i_c shunts part of

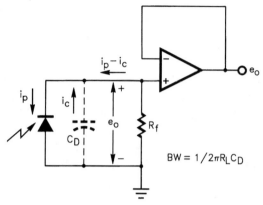

Figure 4.6 Straightforward connection of a photodiode to a load results in output voltage swing across the diode capacitance and a restricted bandwidth.

the diode current away from the load, and this produces a response pole at $f = 1/2\pi R_f C_D$. Typically the large values of R_f used with photodiodes make this pole frequency a severe bandwidth limit. Bootstrap largely removes this limit with the circuit of Fig. 4.7. Fundamentally, only one difference separates this circuit from the previous one. Here, the anode of the diode returns to the amplifier output instead of ground. This voltage follower amplifier monitors the signal voltage at the diode's cathode and drives the anode with the same signal. In the ideal amplifier case, zero voltage remains across the diode and its capacitance. Thus, the bootstrap amplifier removes load

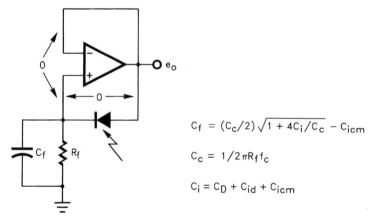

Figure 4.7 Bootstrap drive of a photodiode removes signal voltage from the diode capacitance much like the current-to-voltage converter.

voltage swing from the diode much as the current-to-voltage converter does.

In practice, finite op amp gain again leaves a residual signal voltage on the diode capacitance, just as the earlier current-to-voltage converter did. As before, this residual signal determines the bandwidth for the bootstrap circuit. To quantify this bandwidth, Fig. 4.8 models the basic bootstrap circuit in the manner used previously. In the figure, the amplifier's gain error signal e_o/A_{OL} again appears across the diode. This figure models the photodiode with a current source, shunted by capacitance C_D, and brings out the op amp input capacitances for analysis. With respect to these capacitances, this model differs from the previous Fig. 3.3 current-to-voltage converter. Previously, both C_{id} and C_{icm} remained across the photodiode, adding to its capacitive diversion of signal current. Here, C_{icm} lies across R_f instead, and this difference improves bandwidth, as described below. In return for this improvement, this bootstrap circuit imposes an additional amplifier error source. Here, the bootstrap connection makes the load voltage across R_f a common-mode signal at the op amp inputs. The resulting CMR error of the amplifier typically makes a small addition to the overall circuit error. The earlier current-to-voltage

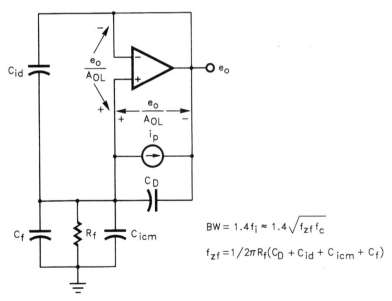

$$BW = 1.4 f_i \approx 1.4 \sqrt{f_{zf} f_c}$$

$$f_{zf} = 1/2\pi R_f (C_D + C_{id} + C_{icm} + C_f)$$

Figure 4.8 Unlike the current-to-voltage converter, the basic bootstrap connection makes the amplifier's C_{icm} input capacitance assist the C_f phase compensation rather than contribute to the bandwidth limit.

converter avoids this error by isolating the amplifier inputs from the load voltage.

4.3.2 Bandwidth analysis of the bootstrap amplifier

Similarities between the current-to-voltage converter and the bootstrap amplifier permit extrapolating the earlier results to this case. Examining the source of the Fig. 4.8 bandwidth limit reveals these similarities. The poles of this circuit result from the amplifier's gain-error signal e_o/A_{OL} that falls across circuit capacitances. Resulting capacitive currents shunt a portion of i_p away from the R_f load, limiting bandwidth. This action replicates that described earlier for the current-to-voltage converter except for one difference in the capacitances that support the e_o/A_{OL} signal. Here, this signal falls across C_D and C_{id}. Previously, the current-to-voltage converter added C_{icm} to this capacitance combination, limiting bandwidth further. Instead, the bootstrap circuit here places C_{icm} in parallel with C_f to aid in the phase compensating bypass of R_f.

Thus, the analysis results for the current-to-voltage converter extrapolate to the Fig. 4.8 bootstrap circuit by accounting for this capacitor difference. First, replacing the sum $C_D + C_{id} + C_{icm}$ of the Chap. 3 results with $C_D + C_{id}$ removes C_{icm} from the net input capacitance C_i. Then, replacing C_f of the earlier results with $C_f + C_{icm}$ introduces the added phase compensation of C_{icm}. The two replacements leave the equation for the op amp's response limit unchanged at

$$f_{pf} = \sqrt{f_{zf} f_c}$$

where $f_{zf} = 1/2\pi R_f(C_D + C_{id} + C_{icm} + C_f)$. However, for a given application, the actual pole frequency increases for the bootstrap circuit. There, C_{icm} supplements C_f's phase compensation, permitting a decrease in the value of C_f. This increase f_{zf} above which, in turn, increases the pole frequency f_{pf}. For small values of diode capacitance C_D, the reduced C_f becomes significant, and the basic bootstrap circuit provides greater bandwidth than the equivalent current-to-voltage converter.

4.3.3 Phase compensating the bootstrap amplifier

As with the current-to-voltage converter of Chap. 3, f_{pf} represents the location of a double pole for the bootstrap circuit. The underlying roll off of the op amp gain produces an inductive characteristic that reacts with circuit capacitances. As frequency increases, the amplifier gain A_{OL} falls and the signal e_o/A_{OL} increases across the circuit capacitances. Simultaneously, the capacitive impedances decline. The combination of a rising signal across a falling impedance again forms the equivalent of an *L-C* tank. Without degeneration, this tank circuit produces oscillation or response ringing, depending upon the Q of the circuit. Degenerating the Q avoids such response disturbances through capacitive bypass of R_f. Together, the amplifier's C_{icm} and the intentionally added C_f bypass this resistor to roll off e_o. This also rolls off the associated e_o/A_{OL} signal reflected to the amplifier input and impressed upon the circuit capacitances. As a result, the bypass counteracts one of the two poles governing the amplifier feedback response to phase compensate the circuit.

Defining a design equation for C_f follows the Chap. 3 analysis of the current-to-voltage converter discussion with two exceptions. First, the bootstrap circuit employs both positive and negative feedback, requiring a redetermination of the feedback factor for feedback analysis. Next, the phase compensation assistance provided by C_{icm} requires adjusting the earlier C_f result to account for the new C_{icm} role here. The discussion that follows briefly summarizes the current-to-voltage converter analysis to accommodate these two changes. Refer to the earlier analysis for greater detail.

Feedback analysis again defines the phase compensation requirement and the resulting bandwidth through the circuit's A_{OL} and $1/\beta$ responses. The bootstrap circuit here alters the circuit's feedback factor β, changing the $1/\beta$ response and the resulting C_f design equation. In Fig. 4.8, both negative and positive feedback appear, and the feedback combination determines the net feedback factor. The fundamental voltage-follower connection of the amplifier makes the negative feedback factor unity, $\beta_- = 1$, through short circuit feedback from the output to the amplifier's inverting input. Positive feedback adds to this through the C_D and C_{id} connections from the output to the noninverting input. Together, these capacitors form a voltage divider with the load circuit, and the voltage divider fraction defines the positive feedback factor

$$\beta_+ = \frac{R_f(C_D + C_{id})s}{1 + R_f(C_D + C_{id} + C_{icm} + C_f)s}$$

The net feedback factor equals the difference between the negative and positive feedback factors,[3] $\beta = \beta_- - \beta_+$, making

$$\beta = \frac{1 + R_f(C_{icm} + C_f)s}{1 + R_f(C_D + C_{id} + C_{icm} + C_f)s}$$

Then, for the bootstrap connection,

$$1/\beta = \frac{1}{\beta_- - \beta_+} = \frac{1 + s/2\pi f_{zf}}{1 + s/2\pi f_{Cf}}$$

where $f_{zf} = 1/2\pi R_f(C_D + C_{id} + C_{icm} + C_f)$ and $f_{Cf} = 1/2\pi R_f(C_f + C_{icm})$. This expression almost repeats the earlier current-to-voltage converter result. As before, this $1/\beta$ result expresses a zero at f_{zf} formed by R_f with the total capacitance of the feedback network, $C_D + C_{id} + C_{icm} + C_f$. However, the counteracting pole at f_{Cf} now benefits from the inclusion of C_{icm} in the phase compensating action of C_f.

Phase compensation requirements guide the C_f selection with the aid of the response curves of Fig. 4.9. There, a rising $1/\beta$ curve approaches the A_{OL} response for a potential 40 dB/decade rate of closure at their intercept. Phase compensation, produced by C_f and C_{icm}, rolls off the $1/\beta$ curve's rise, reducing the rate of closure and restoring stability. However, the $C_f + C_{icm}$ bypass of the load resistor also rolls off bandwidth, discouraging excessive bypass. In compromise, placing the break frequency of the $1/\beta$ roll off at the intercept of the two curves, $f_{Cf} = f_i$, assures 45° of phase margin and optimizes bandwidth. Then, residual gain peaking extends this bandwidth beyond the f_i pole frequency as illustrated by the I-to-V gain curve of the figure. This pole placement sets the bandwidth at BW $= 1.4 f_i = 1.4\sqrt{f_{zf} f_c}$, repeating the earlier result of the current-to-voltage converter. However, as mentioned above, the bootstrap circuit actually improves bandwidth through a decrease in C_f that increases f_{zf} in this BW expression.

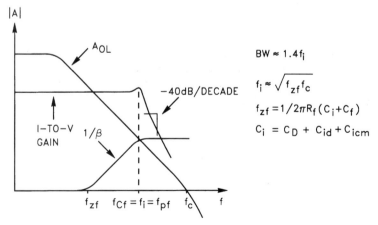

Figure 4.9 The combined $C_f + C_{icm}$ phase compensation of the bootstrap circuit rolls off the circuit's $1/\beta$ response, as before, but the reduced C_f requirement increases f_{zf} to in turn increase the f_i bandwidth limit.

A final comparison with the current-to-voltage converter defines the C_f design equation for the bootstrap circuit. The above $1/\beta$ expression for the bootstrap circuit identifies the phase compensation pole as $f_{Cf} = 1/2\pi R_f(C_f + C_{icm})$. Previously, the current-to-voltage converter produced $f_{Cf} = 1/2\pi R_f C_f$. Phase compensation for both circuits places the f_{Cf} pole at f_i but through the reaction of different capacitances with R_f. As the f_{Cf} equations show, the bootstrap circuit achieves the same compensation as the current-to-voltage converter with a C_f value reduced by an amount equal to C_{icm}. Then, subtracting C_{icm} from the earlier C_f design equation adapts it to the bootstrap case. Previously, a distinction between large and small photodiode capacitances divided the C_f design equation into two cases. However, as described above, the basic bootstrap connection only benefits the small capacitance case and that equation applies here. Then, for the basic bootstrap photodiode amplifier

$$C_f = \frac{C_c}{2} - \sqrt{\frac{1 + 4C_i}{C_c}} - C_{icm}$$

where $C_c = 1/2\pi R_f f_c$ and $C_i = C_D + C_{id} + C_{icm}$. With its minus sign, this equation can yield negative values for C_f. This condition simply indicates that C_{icm} already provides sufficient phase compensation, removing the need for C_f.

4.4 Combining Bootstrapping with the Current-to-Voltage Converter

Even greater bandwidth improvement results from combining the current-to-voltage converter and bootstrap. This combination best serves larger photodiodes where high capacitances preclude the benefit of the basic bootstrap circuit. Together, the current-to-voltage converter and bootstrap introduce two stages of signal removal, leaving very little signal across the diode capacitance. This doubly minimizes capacitive currents that otherwise shunt signal currents to roll off the circuit response. However, effective bootstrapping here requires a separate bootstrap amplifier possessing several demanding characteristics. Fortunately, the bootstrap function relaxes other requirements to make this a practical solution.

4.4.1 The basic combination and its requirements

Figure 4.10 illustrates this bootstrapped I-to-V converter showing the primary op amp of the converter and the unity-gain buffer of the

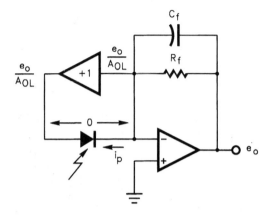

Figure 4.10 Combining bootstrap with the current-to-voltage converter provides a second degree of signal isolation from the diode capacitance that even removes the e_o/A_{OL} residual of the current-to-voltage converter.

bootstrap.[4] This combination removes the bandwidth-limiting signal from the photodiode in two stages as seen from a brief review of earlier discussions. With no signal removal, a photodiode produces an output signal e_o that develops directly across the diode and its capacitance as in the earlier Fig. 4.6. Adding the current-to-voltage converter provides the first stage of signal removal. Here, the converter consists of the op amp and its traditional feedback elements R_f and C_f. As described in Sec. 3.1.2, this basic converter reduces the signal at the cathode of the diode to the gain error signal of the op amp, e_o/A_{OL}. In the basic case, returning the anode of the diode to ground impresses this attenuated signal across the photodiode. There, the large open-loop gain A_{OL} of the op amp dramatically attenuates the signal impressed across the diode. Bootstrap removes even this attenuated signal through the buffer shown here. The unity-gain of the buffer replicates the e_o/A_{OL} signal from the current-to-voltage converter input to the anode of the diode. Then, both terminals of the diode follow the same signal, leaving zero signal across the diode capacitance.

The Fig. 4.10 combination makes photodiode monitoring virtually immune to photodiode capacitance as long as the buffer meets several stringent requirements. The buffer must possess low input capacitance, low noise, wide bandwidth, and low output impedance. For evaluation of these requirements, Fig. 4.11 models the circuit, replacing the buffer with an input capacitance, an output noise source, a bandwidth-limited signal source, and an output resistance. This model again represents the photodiode with a current source and capacitor and also includes the op amp noise voltage e_{ni} with the e_o/A_{OL} input error.

Examining the buffer model elements reveals their importance to the bootstrap function. First, low buffer input capacitance C_{iB} prevents replacing one capacitance problem with another. The buffer removes the error signal e_o/A_{OL} from the diode, avoiding a capacitive current in C_D that otherwise shunts signal current i_p away from feedback resistor R_f. However, capacitance at the buffer input C_{iB} now supports the e_o/A_{OL} signal, producing a new shunting current. To significantly improve bandwidth, the buffer must display an input capacitance much smaller than the diode capacitance, $C_{iB} \ll C_D$.

Similarly, low noise for the buffer avoids exchanging noise problems. With the basic current-to-voltage converter, the input noise voltage of the op amp often dominates noise performance as developed in Chap. 5. This noise adds to the e_o/A_{OL} signal as shown and drives circuit capacitances in the same way as described above for e_o/A_{OL}. For the grounded diode return of the basic converter, input noise voltage e_{ni} drops across the diode capacitance, producing a noise current $e_{ni}C_D s$ that adds to signal current i_p. The unity gain of the buffer removes this noise effect by driving the normally grounded terminal of C_D with the same e_{ni} noise voltage that appears at its other terminal. However, noise introduced by the buffer's signal transmission e_{nB} now appears across C_D, replacing the old noise current with a new one, $e_{nB}C_D s$. To prevent noise degradation, the buffer must assure that $e_{nB} \leq e_{ni}$.

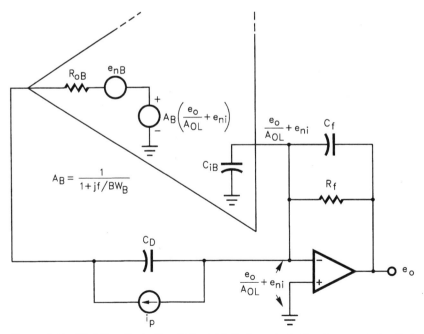

Figure 4.11 Modeling the buffer of Fig. 4.10 displays the buffer characteristics that influence its bootstrap effectiveness.

Wide bandwidth and low output impedance for the buffer assure the accurate signal transmission required for improving the bandwidth of the I-to-V gain. Ideally, the buffer transmits the e_o/A_{OL} error signal from input to output with no attenuation. Any difference between the input and output signals falls across the diode and its capacitance. As described with the basic bootstrap circuit, the bootstrap amplifier produces a gain error signal of its own, and this error signal increases with frequency as the buffer bandwidth rolls off. In the circuit model the $1 + jf/BW_B$ denominator of the buffer gain A_B reflects the buffer's bandwidth effect. Buffer gain roll off due to BW_B introduces an upper limit for the bandwidth improvement of the bootstrapped I-to-V converter. Avoiding this new limit requires making the bandwidth of the buffer much greater than that of the op amp.

However, wide bandwidth alone does not ensure accurate signal transmission through the buffer. This also requires low buffer output impedance to assure that the diode signal current does not develop a significant voltage drop with R_{oB}. Any such drop adds to the gain error signal of the buffer and falls across the diode. There, capacitive current flow in C_D would impose yet another bandwidth limit. A buffer meeting all of the above requirements removes the diode capacitance as a source of bandwidth limitation.

4.4.2 A practical buffer solution

Combined, the above buffer requirements present a formidable design task. Fortunately, the buffer function relaxes other requirements, permitting the use of simple circuits. The buffer does not require the high open-loop gain and low offset voltage required of the current-to-voltage converter's op amp. For that op amp, the very high open-loop gain assures the circuit's response accuracy, and low input offset voltage limits the offset transferred to the circuit output. However, the gain error and offset of the buffer do not directly affect the circuit output. These error signals do appear across the photodiode, but there, significantly larger errors remain acceptable. The buffer's response roll off eventually makes its gain error rise to unacceptable levels, but choosing this buffer for wide bandwidth moves the rise beyond the useful frequency range of the current-to-voltage converter. Thus, low open-loop gain, wide-bandwidth circuits best serve the buffer function by sacrificing gain and offset for bandwidth.

The actual gain error and offset requirements of the buffer remain loose compared to op amp standards. The circuit impresses both error signals across the photodiode, requiring some attention, but only secondary. Buffer gain error allows a small portion of the buffer signal to appear across the diode capacitance. There, it sustains a residual of the high-frequency signal-shunting that previously produced a bandwidth limit. However, the small buffer signal e_o/A_{OL} already represents a greatly attenuated condition, and the residual signal left across the diode here still permits significant bandwidth improvement. For example, even a 10 percent gain error for the buffer permits a 10:1 reduction in the bandwidth limitation imposed by diode capacitance. Similarly, the buffer function accommodates larger offset voltages than most op amp applications. The buffer transfers its offset voltage to the photodiode where it could increase leakage current. However, simple buffer circuits readily limit this offset to tens of millivolts that produce little leakage increase.

Figure 4.12 shows a bootstrapped I-to-V converter with a buffer that addresses the requirements and compromises of the bootstrap function. Basically, this buffer consists of a source follower J_2 biased from current source J_3. The FETs used here limit the input current drawn by the buffer and permit simple realization of the biasing current source. Note that any buffer input current, from the J_2 gate terminal, flows through R_f, producing a component of output offset voltage. The FET J_2 also strongly affects all four of the buffer's critical characteristics: input capacitance, noise, bandwidth,

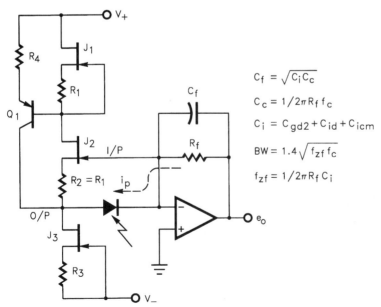

Figure 4.12 Simple buffer circuits provide the best compromise for the wide bandwidth but relaxed gain error and offset requirements of the bootstrap buffer.

and output impedance. Selection of this FET requires attention to gate-drain capacitance, gate-source voltage noise, gain-bandwidth product, and transconductance. With this circuit, the buffer's input capacitance becomes simply $C_{iB} \approx C_{gd2}$, where C_{gd2} is the gate-drain capacitance of J_2. However, the other required buffer properties depend upon the characteristics of multiple circuit elements. Noise and bandwidth for the buffer depend upon a large number of these, preventing simple analysis and favoring evaluation by simulation.

Still, a limited number of circuit elements define the buffer's output impedance. To keep this impedance low, the addition of J_1 and Q_1 provide feedback and loop gain that reduce J_2's normal source-follower output impedance. Without J_1 and Q_1, the i_p current drawn by the diode would flow through R_2 and the source of J_2. Then, the resulting output voltage variation defines the buffer output resistance as $R_{oB} = R_2 + 1/g_{m2}$, where g_{m2} is the transconductance of J_2. Generally, this resistance would develop too great a voltage with i_p, limiting the circuit's bandwidth improvement. This voltage drop falls across the diode, producing capacitive currents that again shunt i_p away from R_f at higher frequencies. Including J_1 and Q_1 introduces feedback that counteracts current changes in R_2 and J_2. With this feedback, load current drawn from the buffer output creates little change in output voltage, i.e., low output resistance. The loop gain driving the feedback starts with J_1. This FET acts as a current source load to the drain of J_2. Any change in J_2's drain current reacts with the high impedance of the J_1 current source to drive the base of Q_1. This latter transistor responds and supplies the current demanded from the buffer output. Essentially the only current change in R_2 and J_2 results from the change in Q_1's base current. Thus, the buffer output resistance drops by a factor approaching the current gain β_1 of Q_1, making $R_{oB} \approx (R_2 + 1/g_{m2})/\beta_1$.

As mentioned, the buffer function permits higher offset voltages than required for the typical op amp. Still, offset must be moderated to avoid excessive diode leakage current. Any offset voltage separating the dc levels of the buffer input and output appears across the diode. Large junction areas make many photodiodes produce high leakage currents under any significant dc bias, and such leakage flows through R_f, producing an output offset voltage as part of e_o. With the buffer shown, simple but compensating bias conditions minimize offset. Source-follower J_2 controls the primary input-to-output signal transfer for the buffer. In this transfer, J_2 introduces an offset component equal to this FET's gate-source voltage V_{GS2}, a negative voltage for the n-channel JFET shown. Adding R_2 produces a counteracting, positive offset shift. However, tolerances in the FET characteristics would tend to make this counteracting effect approximate.

To compensate for this tolerance error, component matching equalizes the dc bias drops of two FETs, J_1 and J_2, and two resistors, R_1 to R_2. In the circuit, J_1 and R_1 combine as an offset-canceling current source that sets the bias current level for J_2 and R_2. As shown, the gate of J_1 connects to the bottom of R_1, establishing a voltage on R_1 equal and opposite that of V_{GS1}. The J_1 current source also sets the drain current for J_2 through the Q_1 feedback described before. From the drain, this current flows through J_2's source and R_2, reproducing the canceling voltage drops described for J_1 and R_1. Here, the same current flows in J_1 and J_2, and making them matched devices produces $V_{GS1} = V_{GS2}$. Similarly, making $R_2 = R_1$ equalizes their voltage drops under the equal current condition described. Ideal component matching reduces the buffer's input-to-output offset voltage to zero. Practical matching reduces it to the tens of millivolts that induce little diode leakage.

4.4.3 Bandwidth analysis of the combination

As described above, bootstrap removes the influence of C_D from the current-to-voltage converter. Then, capacitances remaining at the converter's input define the bandwidth and the phase compensation requirement. The input capacitances of the buffer and op amp present a net input circuit capacitance of $C_i = C_{gd2} + C_{id} + C_{icm}$. Then, the circuit bandwidth follows from the Sec. 3.4 current-to-voltage converter results with C_i replaced by this new expression. For the 45° phase margin underlying previous analyses,

$$\text{BW} = 1.4\sqrt{f_{zf}f_c}$$

where $f_{zf} = 1/2\pi R_f C_i$ and now $C_i = C_{gd2} + C_{id} + C_{icm}$. Here, the reduced C_i extends bandwidth by increasing the frequency $f_{zf} = 1/2\pi R_f C_i$. For large photodiodes, this typically increases bandwidth by a factor of 5 to 10. Note that the C_{gd2} term of C_i simply replaces the C_D term of basic current-to-voltage converter results. Thus, bandwidth improvement with this bootstrapped I-to-V converter requires that $C_{gd2} < C_D$.

This requirement produces a noise-versus-bandwidth compromise in the selection of J_2. The low noise requirement of the buffer encourages the use of a larger FET to achieve a higher g_{m2} and lower associated noise voltage. However, larger FETs also produce larger values for C_{gd2}, limiting the practical fulfillment of the $C_{gd2} < C_D$ requirement. As a result, the bootstrapped I-to-V converter best serves larger photodiodes having larger values of C_D. Smaller photodiode cases benefit more from the basic bootstrap connection of Sec. 4.3.

For the bootstrapped I-to-V converter, the residual C_i capacitance at the op amp input reacts with R_f just like the original C_i of the basic current-to-voltage converter. Chapter 3 describes this reaction and the resulting potential for oscillation. As before, this potential requires phase compensation through a C_f bypass of R_f in Fig. 4.12. However, the reduced input capacitance here reduces the required C_f value. Figure 4.13 guides the phase compensation selection with response curves that replicate those previously described in Sec. 3.3. The reaction of input capacitance with R_f again produces a rise in the $1/\beta$ curve again starting at

$$f_{zf} = 1/2\pi R_f C_i$$

but now $C_i = C_{gd2} + C_{id} + C_{icm}$. The rising $1/\beta$ curve and the falling A_{OL} curve potentially develop a 40-dB rate of closure at their intercept, $f_i = \sqrt{f_{zf}f_c}$, where f_c is the unity-gain crossover of the op amp. That rate of closure would produce oscillation at the frequency f_i except for the effect of the C_f phase compensation. The C_f bypass of R_f rolls off the $1/\beta$ rise, reducing the rate of closure or slope difference at the intercept of the two curves. Choosing C_f to roll off $1/\beta$ right at the intercept restores 45° of phase margin and limits gain peaking to 3 dB as also described in Sec. 3.3.

Selecting C_f for this phase compensation follows from the design equations developed there and two changes to those results adapt them to this case. First, the reduced C_i of this case replaces C_D with

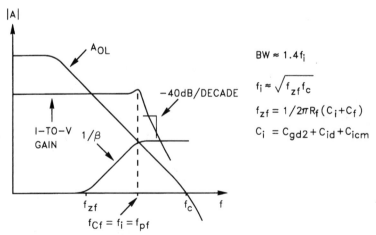

Figure 4.13 Phase compensation of the bootstrapped current-to-voltage converter reproduces familiar response curves, but in practice, the reduced C_i capacitance significantly extends the f_i bandwidth limit.

C_{gd2} as described above. Next, the $C_D \gg C_{\text{gd2}}$ requirement restricts the benefit of this case to larger photodiodes. Thus, the large capacitance version of the previous C_f design equations applies here and for Fig. 4.12,

$$C_f = \sqrt{C_i C_c}$$

where $C_c = 1/2\pi R_f f_c$, and $C_i = C_{\text{gd2}} + C_{\text{id}} + C_{\text{icm}}$.

References

1. J. Graeme, "Phase Compensation Optimizes Photodiode Bandwidth," *EDN*, May 7, 1992, p. 177.
2. J. Graeme, "Circuit Options Boost Photodiode Bandwidth," *EDN*, May 21, 1992, p. 155.
3. J. Graeme, "Feedback Models Reduce Op Amp Circuits to Voltage Dividers," *EDN*, June 20, 1991, p. 139.
4. O. Compastro, "Utilizacion de Fotodetectores de Gran Area en Sistemas de Gran Ancho de Banda," *Revista Telegrafica Electronica,* July, 1984, p. 832.

Chapter 5

Noise

As a photodiode amplifier, the current-to-voltage converter exhibits a complex noise behavior. Basic noise components result from the circuit's feedback resistor and the amplifier input's noise current and noise voltage. The resistor and current noise sources produce output noise components that follow from intuition. However, the amplifier's input noise voltage receives an unexpected high-frequency gain. This gain results from the circuit's combination of a high feedback resistance and the diode capacitance at the amplifier input. Parasitic capacitance around the feedback resistor and the characteristic $1/f$ response of the noise voltage further complicate the circuit's final noise response. Multiple poles and zeros shape this response, posing a formidable noise analysis task. However, breaking the analysis into a series of frequency regions, separated by the poles and zeros, restores simplicity. This regional analysis also permits comparison of relative noise effects to identify circuit changes that optimize performance. Following the individual analyses of these resistor, noise current, and noise voltage effects, a summarized listing of noise equations permits step-by-step analysis of specific applications. Evaluation of these results as a function of feedback resistance shows that the dominant noise source of a photodiode amplifier changes from the op amp to the resistor and then to the combination.

5.1 General Noise Effects

The simple appearance of the Fig. 5.1 photodiode amplifier deceivingly suggests a simple noise performance. There, the photodiode operates at zero bias and presents a high impedance to the op amp input. This suggests an op amp with a unity feedback factor produced by a simple resistor feedback. As such, the circuit would normally produce a noise

88 Chapter Five

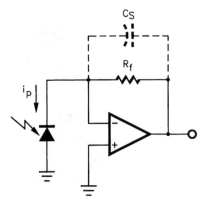

Figure 5.1 At first, the simple structure of the photodiode amplifier suggests simple noise performance.

gain of unity and transfer the op amp's input noise voltage directly to the circuit output. However, the high impedance of the photodiode is relative and variable. Photodiode amplifiers also make R_f a high impedance, and photodiodes typically exhibit large shunt capacitances that roll off the diode impedance. This combination produces noise gain peaking at higher frequency and a complex response to the op amp's input noise voltage. In addition, the high R_f resistance increases the significances of that resistor's noise and the noise of the amplifier input current. For each of the circuit's noise sources, complete noise evaluation begins with identifying the corresponding spectral noise densities and noise gains. Then, root mean square (rms) integration incorporates frequency response effects for each of the three.

5.1.1 Noise densities and noise gains

The photodiode amplifier's high-value feedback resistor increases the circuit's noise, both directly and indirectly. The discussion below first examines the resistor's effects upon the spectral noise densities and a later discussion examines the total rms noise produced at the circuit output. As a noise measure, spectral noise density reflects the noise contained in a 1-Hz bandwidth at a given frequency, and this density may vary with frequency. RMS integration sums the noise-density effects at the individual frequencies, reflecting the total noise effect of a given noise source. For the photodiode amplifier, the spectral noise densities e_{noR}, e_{noi}, and e_{noe} represent the output noise components produced by the resistor itself, the amplifier's input current noise, and the amplifier's input voltage noise. High feedback resistance increases all three output noise components. These noise sources combine to produce the net, output noise density e_{no}. As will be described, the e_{noR} and e_{noi} noise densities remain constant with frequency until a bandwidth limit truncates their effects. However, the e_{noe} noise density varies with frequency and rolls off at a different bandwidth limit.

Figure 5.2 models the basic photodiode amplifier of Fig. 5.1 for noise analysis. In the model, i_p, R_D, and C_D represent the photodiode. Capacitance C_{ia} and noise sources i_{ni} and e_{ni} represent the relevant characteristics of the amplifier input. Finally, noise source e_{nR} represents the noise voltage of the feedback resistor. The noise contributed directly by the feedback resistor has a spectral density of[1] $e_{nR} = \sqrt{4KTR_f}$. Here, K is Boltzman's constant, 1.38×10^{-23} J/°K, and T is temperature in degrees Kelvin, °C + 273. This noise voltage transfers to the output of a current-to-voltage converter without amplification, making $e_{noR} = e_{nR}$. Increasing the size of R_f raises $e_{noR} = \sqrt{4KTR_f}$ by a square root relationship but also increases the desired output signal $i_p R_f$ by a direct proportionality. Thus, the resulting signal-to-noise ratio, $i_p R_f / e_{noR} = i_p \sqrt{R_f/4KT}$, improves by the square root of the resistance R_f. This suggests maximizing R_f to

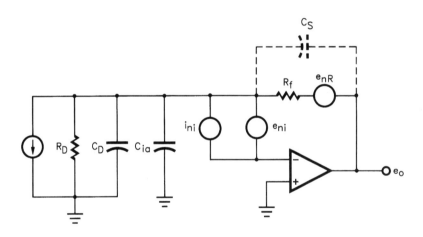

$$e_o = i_p R_f + e_{no} \qquad e_{no} = \sqrt{(e_{noR})^2 + (e_{noi})^2 + (e_{noe})^2}$$

$$e_{noR} = \sqrt{4KTR_f} \qquad e_{noi} = R_f\sqrt{2qI_B-} \qquad e_{noe} = \frac{(1 + R_f C_i s)}{(1 + R_f C_s s)} e_{ni}$$

$$C_i = C_D + C_{ia}, \quad C_{ia} = C_{id} + C_{icm}$$

Figure 5.2 The noise analysis model of the basic photodiode amplifier breaks the photodiode into its current and impedance components, separates capacitances and noise sources from the amplifier input, and adds the noise voltage of the feedback resistor.

optimize the signal-to-noise ratio. However, this signal-to-noise relationship only includes the noise produced by the feedback resistor, and increasing R_f potentially decreases bandwidth. Depending upon the circuit elements used, other noise effects may dominate noise performance as described later in Sec. 5.3.2. There, an evaluation of noise dominance versus feedback resistance identifies the resistance range where increasing R_f improves the overall signal-to-noise ratio.

Indirectly, the R_f feedback resistor also influences the current-to-voltage converter's noise through interaction with the op amp's noise sources. In the model, i_{ni} and e_{ni} represent the amplifier's input noise current and input noise voltage. Noise current i_{ni} represents the shot noise of the input bias current I_{B-} and has a noise density of[1] $i_{ni} = \sqrt{2qI_{B-}}$, where q is the charge on an electron, 1.6×10^{-19} C. This noise current flows directly through the feedback resistor, producing a noise voltage of $e_{nRi} = i_{ni}R_f = R_f\sqrt{2qI_{B-}}$. Like the noise voltage of R_f itself, this new noise voltage transfers to the circuit output with unity gain and $e_{noi} = e_{nRi}$. Choosing an op amp with an I_{B-} in the picoamp range generally makes this noise component negligible for practical levels of feedback resistance.

5.1.2 Noise gain peaking

In a more complex relationship, the amplifier's input noise voltage e_{ni} receives an amplification characterized by a noise gain peaking. At first, it would seem that e_{ni} should also transfer to the circuit output with low gain, but capacitances modify this gain. In the figure, C_D, C_{ia}, and C_S represent the capacitances of the photodiode, the amplifier input, and a stray capacitance, respectively. At dc, the circuit presents a noise voltage gain of $A_{ne} = 1 + R_f/R_D$ to e_{ni}. There, the very large diode resistance R_D keeps this voltage gain near unity. However, at higher frequencies, C_D bypasses R_D, shunting its high resistance and increasing A_{ne}. Amplifier capacitance C_{ia} appears in parallel with C_D, adding to this effect. Here, $C_{ia} = C_{id} + C_{icm}$, where C_{id} is the differential input capacitance between the amplifier inputs and C_{icm} is the common-mode input capacitance from the inverting amplifier input to ground. Together, the diode and amplifier capacitances produce a net input capacitance $C_i = C_D + C_{id} + C_{icm}$. This capacitance shunts R_D, converting the noise gain received by e_{ni} to $A_{ne} = 1 + R_f/[R_D \| (1/C_i s)]$. At higher frequencies, this gain becomes $A_{ne} \approx 1 + R_f C_i s$ and rises in direct proportion to frequency. Frequently, the demand for high signal gain makes both C_D and R_f large, initiating this rising noise gain at fairly low frequencies. However, large R_f values also introduce a limit to the ac gain rise. There, stray capacitance shunting of R_f, shown by the dashed C_S, counteracts the gain rise produced by C_i.

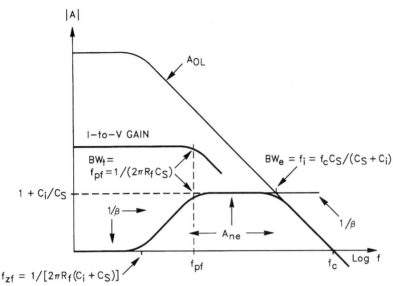

Figure 5.3 High feedback resistance and the capacitance of the input circuit make the photodiode amplifier's noise gain A_{ne} rise at higher frequency until leveled by stray capacitance and finally rolled off by the amplifier open-loop response.

Figure 5.3 illustrates the various capacitance effects through the noise gain curve A_{ne}, plotted with op amp gain-magnitude curve A_{OL}. Two other curves of the figure represent the circuit's current-to-voltage gain and $1/\beta$ response. Prior to the $1/\beta$ intercept with A_{OL}, $A_{ne} = 1/\beta$. The A_{ne} curve begins with the unity gain first expected in the above circuit discussion and then experiences a response zero at $f_{zf} = 1/2\pi R_f(C_i + C_S)$. There, feedback shunting due to the reaction of R_f and C_i initiates an A_{ne} rise. Stray capacitance shunting of R_f terminates this rise with a pole at $f_{pf} = 1/2\pi R_f C_S$, leveling the noise gain at a plateau level of $1 + C_i/C_S$. With large-area photodiodes, a high-value C_D makes C_i hundreds of picofarads, producing a gain plateau in the hundreds as well. That gain continues to higher frequencies until rolled off by the op amp's A_{OL} bandwidth limit. Within the A_{OL} response boundary of the op amp, the A_{ne} response makes the output noise density due to e_{ni}

$$e_{noe} = \frac{1 + R_f(C_i + C_S)s}{1 + R_f C_S s} e_{ni}$$

Here, a zero and a pole mark the noise gain effects of C_i and C_S. Roll off of the amplifier's A_{OL} response adds a second pole, described later but temporarily omitted here. Increasing R_f above, to increase the current-to-voltage gain, moves the zero and pole of e_{noe} to lower frequencies, encompassing a greater frequency spectrum with high noise gain. Thus, the current-to-voltage converter as a photodiode amplifier introduces a very serious gain-versus-noise compromise.

The preceding discussions describe the spectral noise densities that three noise sources create at the current-to-voltage converter's output. However, the above e_{noR}, e_{noi}, and e_{noe} represent individual noise contributions rather than the circuit's total noise. Combining these individual contributions requires rms rather than linear summation due to the uncorrelated nature of the noise sources. This rms summation defines the spectral noise density at the current-to-voltage converter's output as

$$e_{no} = \sqrt{e_{noR}^2 + e_{noi}^2 + e_{noe}^2}$$

where $e_{noR} = \sqrt{4KTR_f}$, $e_{noi} = R_f\sqrt{2qI_{B-}}$, and $e_{noe} = \{[1+R_f(C_i + C_S)s]/[1+R_fC_Ss]\}e_{ni}$. Still, this is not the final rms summation that yields the circuits total output noise. The above summation merely combines three noise density results, producing the net noise density e_{no}. This noise density remains a function of frequency, requiring further rms analysis to define total output noise. RMS integration of noise density over frequency defines this total noise as described below.

5.1.3 Total rms output noise

The spectral nature of the above noise densities restricts numerical noise analysis to one frequency at a time. Broadband evaluation of output noise requires summing the individual frequency effects over the bandwidth of the amplifier. Simple rms integration performs this summation, yielding noise analysis equations for the photodiode amplifier. Gain and bandwidth affect the three fundamental noise sources differently and first require separate rms analyses of individual output noise effects. Then, rms combination of the three separate effects defines the total output noise. In this process, the separate analyses permit identification of the predominant noise effect in a given application. The discussion below first considers the simpler effects of e_{noR} and e_{noi}. A later examination defines the differences that extend this process to the third noise component e_{noe}.

RMS analysis summarizes the effect of a given noise source by mathematically integrating its spectral noise effect over frequency. This integration combines the circuit's noise density, gain, and bandwidth characteristics into a single noise indicator, the RMS value of the total noise. For each noise source, converting noise density to RMS noise requires evaluating the integral[1,2]

$$E_{no}^2 = \int_0^\infty |A_n e_n|^2 df$$

Here, e_n represents an input-referred noise source, and A_n represents the corresponding noise gain supplied to e_n. Noise gain A_n includes the relevant bandwidth characteristic.

The constant noise gains and constant noise densities associated with the circuit's feedback resistor and the amplifier's input current simplify the integration for these two noise sources. There, constant noise gains of $A_n = 1$ ideally transfer the effects of these two noise sources directly to the circuit output, developing the output noise densities e_{noR} and e_{noi} described before. However, the bandwidth limit of the circuit's transresistance BW_t rolls off this ideal transfer response at some higher frequency. For high-gain current-to-voltage converters, this bandwidth limit occurs where the stray capacitance C_S breaks with R_f. In other cases, the addition of phase compensation capacitance in parallel with C_S modifies this break frequency. In either case, this break frequency simultaneously rolls off the output signals developed by the photodiode current, the resistor noise, and the amplifier's current noise. Then, $A_n = 1/(1 + jf/BW_t)$ where $BW_t = 1/2\pi R_f C_S$, and this defines one variable of the above integral. The other variable, e_n, follows from the analyses of Sec. 5.1.1 for the resistor and current noise effects. There, $e_{nR} = \sqrt{4KTR_f}$ and $e_{nRi} = R_f\sqrt{2qI_{B-}}$ describe the constant noise densities produced by the feedback resistor and amplifier input noise current. Evaluating the rms integral for these A_n and e_n conditions defines the rms output noise contributions as

$$E_{noR} = \sqrt{2KTR_f \pi BW_t}$$

and

$$E_{noi} = R_f\sqrt{q \pi BW_t I_{B-}}$$

where $BW_t = 1/2\pi R_f C_S$.

Above, the BW_t expression assumes the high-gain photodiode amplifier case where the bypass of stray capacitance automatically defines $BW_t = 1/2\pi R_f C_S$. In other cases, wider bandwidth photodiode amplifiers add a feedback bypass capacitor C_f to control stability and this capacitance adds to C_S in setting BW_t. Then, the total feedback bypass capacitance $C_f + C_S$ makes $BW_t = 1/2\pi R_f(C_f + C_S)$. With this modification, the capacitive shunting described and the noise results expressed by the E_{noR} and E_{noi} equations still apply.

5.2 The Effect of Op Amp Input Noise Voltage

Constant noise densities, constant noise gains, and a common noise bandwidth simplified the rms integration of the e_{nR} and e_{nRi} noise voltages above. The e_{ni} noise voltage, however, displays a variable noise density, experiences a frequency-dependent amplification, and receives a greater noise bandwidth than e_{nR} and e_{nRi}. These differences require a more detailed description of $e_{noe} = A_{ne}e_{ni}$ before performing the corresponding integration. First, an intuitive examination of these variables produces an expression for the circuit's output noise density resulting from e_{ni}. This expression identifies response poles and zeroes that mark the endpoints of five distinct e_{noe} regions. Considering each region separately greatly simplifies the required rms integration and produces five output noise components. A later rms summation defines the net noise effect.

5.2.1 An intuitive derivation of the e_{noe} noise component

Amplifier noise voltage e_{ni} exhibits a characteristic $1/f$ response at lower frequencies and the corresponding noise gain of the circuit A_{ne} increases at higher frequencies. Also, the stray capacitance bypass of R_f, which set bandwidth BW_t above, levels but does not roll off A_{ne}. Only the open-loop response of the op amp finally rolls off this noise gain. As a result, the $e_{noe} = A_{ne}e_{ni}$ product of the rms integral becomes a complicated function of frequency. However, two intuitive modifications to the earlier e_{noe} result ease the derivation of the e_{noe} characteristic equation. First, the addition of a response pole accounts for the roll off effect of the op amp, and then replacing e_{ni} with its characteristic equation adds the $1/f$ effect.

For the first modification, consider the frequency-dependent noise gain A_{ne} of Fig. 5.3. As frequency increases, this gain first rises, due to the circuit's input capacitance, and then levels, due to capacitance bypass of the feedback resistor. From Sec. 5.1.2 this produces

$$e_{noe} = \frac{1 + R_f(C_i + C_S)s}{1 + R_f C_S s} e_{ni}$$

prior to the $1/\beta$ intercept with A_{OL}. This intercept modifies the e_{noe} response, adding a pole at f_i and making

$$e_{noe} = A_{ne}e_{ni} = \frac{1 + R_f(C_i + C_S)s}{(1 + R_f C_S s)(1 + s/2\pi f_i)} e_{ni}$$

Here, $f_i = \beta_H f_c$, where β_H represents the high-frequency value of the

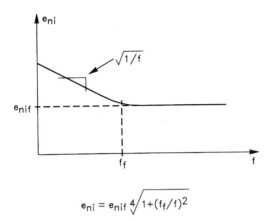

Figure 5.4 Op amp input noise voltage e_{ni} displays a $1/f$ region at low frequencies followed by a noise floor level e_{nif} beginning at the $1/f$ corner frequency f_f.

circuit's feedback factor and f_c represents the unity-gain crossover of the op amp. For the high-gain case, circuit capacitances make $\beta_H = C_S/(C_i + C_S)$ for $f_i = f_c C_S/(C_i + C_S)$.

Including the frequency dependence of e_{ni} in the above equation completes the e_{noe} derivation. Figure 5.4 illustrates this dependence with the characteristic $1/f$ response at lower frequencies. There, the e_{ni} response displays a decreasing magnitude up to the noise-floor corner at f_f. In this initial region, noise power declines in proportion to $1/f$ and, thus, noise voltage e_{ni} declines in proportion to $\sqrt{1/f}$. After the f_f corner, the e_{ni} response follows its constant floor level e_{nif}. The expression $e_{ni} = e_{nif}\sqrt{(\omega_f + s)/s}$, where $\omega_f = 2\pi f_f$, describes this noise-density response. Reducing this expression to the magnitude response $e_{ni} = e_{nif}\sqrt[4]{1 + (f_f/f)^2}$ demonstrates the correspondence to the figure. For higher frequencies, where $f \gg f_f$, this magnitude expression produces the noise-density floor level $e_{ni} = e_{nif}$. For lower frequencies, $f \ll f_f$, this expression produces the expected $\sqrt{1/f}$ proportionality in $e_{ni} = e_{nif}\sqrt{f_f/f}$. Finally, at the corner frequency the expression produces $e_{ni} = e_{nif}\sqrt[4]{2}$. There, the power-to-voltage conversion of the $1/f$ analysis produces the $\sqrt[4]{2}$ term of a square rooted response zero. Substituting this e_{ni} expression into the preceding e_{noe} equation yields

$$e_{noe} = A_{ne}e_{ni} = \frac{[1 + R_f(C_i + C_S)s]\sqrt{(\omega_f + s)/s}}{(1 + R_f C_S s)(1 + s/\omega_i)} e_{nif}$$

5.2.2 Simplifying the E_{noe} analysis

Converting the above noise density into an rms noise result again requires evaluating the integral

$$E_{noe}^2 = \int_0^\infty |A_{ne}e_{ni}|^2 \, df = \int_0^\infty |e_{noe}|^2 df$$

96 Chapter Five

Just examining the preceding e_{noe} expression suggests the very complex result that this integration produces. However, separating the integration into five response segments reduces the integration task to a more manageable and more insightful evaluation.[3] Then, approximation segments permit manual analysis and reveal the relative noise effects of different frequency ranges. Knowing these relative effects and their circuit origins permits design changes that optimize noise performance.

Figure 5.5 shows the frequency response of e_{noe} and the analysis approximation regions. Due to the logarithmic scales of response plots, linear addition of the preceding A_{ne} and e_{ni} curves of Figs. 5.3 and 5.4 produces the $e_{noe} = A_{ne}e_{ni}$ curve shown here. This summation reproduces previously identified response singularities at the boundaries of distinct response regions. Frequency f_f marks the $1/f$ noise corner, f_{zf} and f_{pf} mark the zero and pole of the noise gain peaking, and f_i marks the response roll off produced by the $1/\beta$ intercept. Between each of these frequencies, e_{noe} displays a fairly consistent response slope that eases the rms integration. For this integration, different straight-line approximations will represent e_{noe} in each of the five regions shown. In each region, these approximations follow the region's predominate slope.

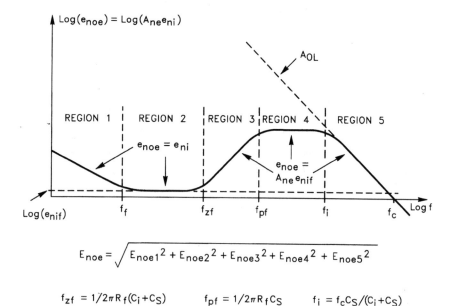

Figure 5.5 Combining the A_{ne} response of Fig. 5.3 with the e_{ni} response of Fig. 5.4 produces the $e_{noe} = A_{ne}e_{ni}$ response here with five distinct regions that separate the rms integration into simpler segments.

Variations in both A_{ne} and e_{ni} contribute to the characteristic shape of the e_{noe} curve, but fortunately, the two variations do not occur simultaneously in any given frequency region. In regions 1 and 2, e_{noe} follows the e_{ni} response of Fig. 5.4. There, a constant $A_{ne} = 1$ makes $e_{noe} = e_{ni}$, transferring e_{ni} directly to the circuit output. Gain peaking interrupts this direct transfer beginning in region 3 and extending through region 5. There, the A_{ne} peaking response of Fig. 5.3 produces a similar peaking for e_{noe} in Fig. 5.5. However, in these last three regions, e_{ni} remains constant at its noise floor level of e_{nif}, making $e_{noe} = A_{ne} e_{nif}$.

5.2.3 The E_{noe} components

Performing the rms integration for the five regions of Fig. 5.5 produces noise components E_{noe1} through E_{noe5} for later rms combination into a net E_{noe} result. This combination defines the total output noise E_{noe}, created at the photodiode amplifier's output by amplifier input noise voltage e_{ni}. In regions 1 and 2 of the figure, e_{noe} replicates the amplifier input noise voltage e_{ni}. In region 1, e_{noe} follows the $\sqrt{1/f}$ slope of the e_{ni} response up to the f_f corner frequency. There, $e_{noe} = e_{ni} \approx e_{nif}\sqrt{f_f/f}$, as described earlier in Sec. 5.2.1. Application of the rms integration to this e_{noe} expression produces the rms noise expression

$$E_{noe1}^2 = \int_0^{f_f} \left| \frac{e_{nif}^2 f_f}{f} \right| df = e_{nif}^2 f_f \ln \frac{f_f}{0}$$

This result displays the common dilemma of $1/f$ noise theory. The 0 limit of the integral transfers to the denominator of the result, rendering the latter indeterminate. Indeed, a noise magnitude proportional to $\sqrt{1/f}$ approaches infinity as f approaches zero. This dilemma has yet to be resolved in the theoretical analysis of $1/f$ noise.[4]

In practice, replacing the 0 limit with a nonzero, low-frequency f_1 restores the analysis and delivers results in good agreement with measurements. Typically, making $f_1 = 0.01$ Hz instead of 0 Hz suffices and only neglects a very small portion of the frequency spectrum of integration. Then,

$$E_{noe1}^2 = \int_{f_1}^{f_f} \left| \frac{e_{nif}^2 f_f}{f} \right| df = e_{nif}^2 f_f \ln \frac{f_f}{f_1}$$

making the rms noise expression for region 1

$$E_{noe1} = e_{nif} \sqrt{f_f \ln \frac{f_f}{f_1}}$$

where e_{nif} is the e_{ni} noise floor level and $f_1 \approx 0.01$ Hz.

In region 2, the combination of constant noise gain and constant noise density simplifies the rms analysis. There, the circuit's unity feedback factor continues a noise gain of $A_{ne} = 1$ from f_f up to the f_{zf} transition that begins the next analysis region. Also in region 2, the amplifier's input noise voltage follows its noise floor at a level of e_{nif}. Then, for region 2

$$E_{noe2}^2 = \int_{f_f}^{f_{zf}} |e_{nif}|^2 \, df = e_{nif}^2 (f_{zf} - f_f)$$

and

$$E_{noe2} = e_{nif}\sqrt{f_{zf} - f_f}$$

where $f_{zf} = 1/2\pi R_f(C_i + C_S)$, and $C_i = C_D + C_{id} + C_{icm}$.

Regions 3 through 5 of Fig. 5.5 display the noise gain peaking that amplifies the op amp input noise voltage, e_{ni}. In each of these regions, e_{ni} remains at its noise floor level of $e_{ni} = e_{nif}$. Region 3 begins the A_{ne} gain peaking with the zero at f_{zf}, making e_{noe} rise with the 1:1 slope of a single-zero response. There, A_{ne} increases from its unity level in direct proportion to frequency, making $A_{ne} = f/f_{zf}$ and $e_{noe} = A_{ne}e_{nif} = e_{nif}f/f_{zf}$. The rising e_{noe} response of this region continues up to the frequency f_{pf}. Thus,

$$E_{noe3}^2 = \int_{f_{zf}}^{f_{pf}} \left|\frac{e_{nif}f}{f_{zf}}\right|^2 df = \left(\frac{e_{nif}}{f_{zf}}\right)^2 \frac{(f_{pf}^3 - f_{zf}^3)}{3}$$

and

$$E_{noe3} = \frac{e_{nif}}{f_{zf}}\sqrt{\frac{f_{pf}^3 - f_{zf}^3}{3}}$$

where $f_{pf} = 1/2\pi R_f C_S$, $f_{zf} = 1/2\pi R_f(C_i + C_S)$, and $C_i = C_D + C_{id} + C_{icm}$.

Following f_{pf}, the e_{noe} response levels in a plateau where stray capacitance limits the noise gain to $A_{ne} = 1 + C_i/C_S$. In this region 4, the amplifier's input noise voltage continues at $e_{ni} = e_{nif}$, making $A_{ne}e_{ni} = (1 + C_i/C_S)e_{nif}$. This region extends up to the f_i intercept and

$$E_{noe4}^2 = \int_{f_{pf}}^{f_i} \left|\left(\frac{1+C_i}{C_S}\right)e_{nif}\right|^2 df = \left[\left(\frac{1+C_i}{C_S}\right)e_{nif}\right]^2 (f_i - f_{pf})$$

for

$$E_{noe4} = \left(\frac{1+C_i}{C_S}\right)e_{nif}\sqrt{(f_i - f_{pf})}$$

where $f_i = f_c C_S/(C_i + C_S)$ and $f_{pf} = 1/2\pi R_f C_S$ from before.

Region 5 concludes the e_{noe} response by following the A_{OL} roll off from f_i out to infinite frequency. This roll off follows a single-pole response up to the amplifier's unity-gain crossover frequency f_c and there, $A_{ne} = 1$. Between f_i and f_c, the 1:1 slope of the amplifier's single-pole response makes $A_{ne} = f_c/f$. Also in this region, the amplifier's input noise voltage continues at its constant floor level of $e_{ni} = e_{nif}$, making $e_{noe} = e_{nif}f_c/f$. After f_c, other amplifier poles alter the roll off, but there, the low and declining level of e_{noe} minimizes their effects upon the rms noise of the region. Neglecting these added poles, the region 5 rms noise integral becomes

$$E_{noe5}^2 = \int_{f_i}^{\infty} \left|\frac{e_{nif}f_c}{f}\right|^2 df = \frac{(e_{nif}f_c)^2}{f_i}$$

and

$$E_{noe5} = (e_{nif}f_c)\sqrt{\frac{1}{f_i}}$$

5.3 Combining the Noise Effects

RMS summations combine the seven noise components of the preceding sections to define the photodiode amplifier's net output noise. Evaluations of these summations define two parameters of noise dominance. The first evaluation examines frequency dependence and shows that the higher-frequency regions of the e_{noe} response dominate the noise contribution of e_{ni}. The second examines resistor dependence and shows that the noise dominance varies between the op amp, the feedback resistor, and the combination.

5.3.1 The noise analysis summary

Two rms voltage summations conclude the noise analysis for the photodiode amplifier. The first combines the five components of E_{noe} noise initially separated in the Sec. 5.2.3 analyses. The second combines the net E_{noe} result with the E_{noR} and E_{noi} components described in Sec. 5.1.3. At first, it might be expected that the E_{noe} noise components, all from the same noise source, should sum linearly to produce the net E_{noe} result. Also, the integration function of the preceding rms analyses suggests linear summation for combining the noise contributions at different frequencies into an rms result. However, closer physical and mathematical examinations of these noise components indicate the need for rms summation instead. While the five components result from the same noise source, nothing in the random physical behavior of that source gives rise to correlation between the noise in one frequency band with that in any other. Uncorrelated noise components such as these add in an rms fashion. Further, the linear summation of the rms integration

$$E_{\text{noe}}^2 = \int_0^\infty |e_{\text{noe}}|^2 \, df$$

operates upon the square of the noise voltage, and the final E_{noe} equals the square root of the integral result. Thus, combining the individual results produces the square root of the sum of the squares of an rms summation.

Combining the five components of E_{noe} in this manner defines the net rms noise produced by op amp input noise voltage at the current-to-voltage converter output as

$$E_{\text{noe}} = \sqrt{E_{\text{noe1}}^2 + E_{\text{noe2}}^2 + E_{\text{noe3}}^2 + E_{\text{noe4}}^2 + E_{\text{noe5}}^2}$$

For a given application, numerical evaluation of the E_{noe} expression reveals the relative significance's of E_{noe1} through E_{noe5}. Different circuit factors such as e_{nif}, C_i, and f_c influence these noise voltage segments. Thus, breaking the rms integration into these five components provides insight into the circuit factors requiring attention for noise reduction. Typically, E_{noe4} and E_{noe5} dominate the E_{noe} result due to the greater frequency spectrums encompassed by regions 4 and 5. Visual evaluation of Fig. 5.5's five regions fails to reveal the degree of this dominance due to the logarithmic compression of the curves two axes.

Rescaling these axes for linear rather than logarithmic representation reveals the relative significance of each region in Fig. 5.6. While uncommon for frequency response plots, this linear scaling reexpands the magnitude and frequency axes, dramatically improving visual perception of relative effects. For the five e_{noe} regions, the physical area under the curve now directly represents the relative noise contribution of a given region. This area drops to insignificance for regions 1 through 3. Region 4 retains a significant area primarily due to its high noise magnitude. Then, region 5 dominates the undercurve area with an expanse that covers most of graph's extent. Thus, to minimize the noise effect of e_{ni}, generally focus upon circuit factors that affect regions 4 and 5. However, this generalization does not necessarily hold for low-frequency applications of photodiodes. There, subsequent filtering of the photodiode amplifier output often removes the noise contribution of regions 4 and 5, returning the contributions of regions 1 through 3 to prominence.

Combining the above E_{noe} result with the E_{noR} and E_{noi} results of Sec. 5.1.3 produces the total output noise E_{no} of the photodiode ampli-

Noise   101

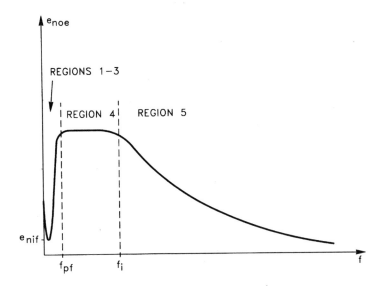

NOTE: Graph axes follow linear rather than Log scales.

Figure 5.6 Rescaling Fig. 5.5 with linear rather than logarithmic axes provides visual perception of the true significance of each noise region through the relative areas enclosed under the e_{noe} curve.

fier. Once again, rms summation combines the noise components. To simplify computation of a specific E_{no} result, the listing below summarizes the seven fundamental noise components described before. The equations below depend upon other expressions that define noise magnitudes, cutoff frequencies, and capacitances. For simplicity, the listing defines these other expressions only at their first occurrence. Step-by-step calculation of E_{no} for the basic current-to-voltage converter proceeds through the individual calculations:

$E_{\text{noe1}} = e_{\text{nif}}\sqrt{f_f \ln(f_f/f_1)}$, where e_{nif} is the e_{ni} noise floor level, and f_f is the $1/f$ noise corner frequency. As explained in Sec. 5.2.3, setting $f_1 \approx 0.01$ Hz avoids a theoretical contradiction and ensures good analytical accuracy.

$E_{\text{noe2}} = e_{\text{nif}}\sqrt{f_{\text{zf}} - f_f}$, where $f_{\text{zf}} = 1/2\pi R_f(C_i + C_S)$, and $C_i = C_D + C_{\text{id}} + C_{\text{icm}}$. For applications requiring the phase compensation element C_f, replace C_S with $C_f + C_S$ here and in the steps that follow. Here and in the E_{noe4} analysis, numerical evaluation can produce the square root of a negative number. Then, substitute zero for the result as explained below.

$E_{noe3} = (e_{nif}/f_{zf})\sqrt{(f_{pf}^3 - f_{zf}^3)/3}$, where $f_{pf} = 1/2\pi R_f C_S$, and C_S is the stray feedback capacitance.

$E_{noe4} = (1 + C_i/C_S)e_{nif}\sqrt{f_i - f_{pf}}$, where $f_i = f_c C_S/(C_i + C_S)$.

$E_{noe5} = (e_{nif} f_c)\sqrt{1/f_i}$, where f_c is the op amp's unity-gain crossover frequency.

$E_{noe} = \sqrt{E_{noe1}^2 + E_{noe2}^2 + E_{noe3}^2 + E_{noe4}^2 + E_{noe5}^2}$, the total rms output noise resulting from the noise voltage density e_{ni} of the amplifier input.

$E_{noR} = \sqrt{2KTR_f \pi BW_t}$, where $BW_t = 1/2\pi R_f C_S$. For applications that add a feedback capacitance C_f, replace C_S with $C_f + C_S$.

$E_{noi} = R_f\sqrt{q \pi BW_t I_{B-}}$.

$E_{no} = \sqrt{E_{noR}^2 + E_{noi}^2 + E_{noe}^2}$.

Special cases potentially affect the E_{noe2} and E_{noe4} results. These noise components cover flat response regions in Fig. 5.5 that certain circuit combinations eliminate. In that figure, the condition $f_{zf} < f_f$ eliminates region 2, and $f_i < f_{pf}$ eliminates region 4. Very large R_f, C_D time constants can produce one or both of these conditions. Then, the eliminated region or regions produce zero noise contribution. However, to cover all possible cases, the segmented E_{noe} analysis must include the potential noise contributions of regions 2 and 4. Numerical evaluations of the E_{noe2} and E_{noe4} equations easily identify these special cases by delivering the square root of a negative number. Then, simply substitute zero for the region's numerical result. When this results for E_{noe4}, it signals the potential need for the phase compensation described in Chap. 3.

5.3.2 Identifying the predominant noise effect

For a given application, the numerical evaluation of E_{no}, following the listing above, identifies the relative significances of the individual noise contributors. There, numeric comparison of terms in both the E_{noe} and E_{no} calculations reveals the actual factors that significantly affect noise performance. Without these comparisons, traditional intuition alone leads to suboptimal design choices. For example, traditional intuition biases the op amp selection toward low input noise voltage. However, this bias risks unnecessarily sacrificing other op amp characteristics. By itself, the amplifier's noise voltage only dominates noise performance in a portion of photodiode amplifier designs. In other cases, noise from the feedback resistor dominates, making the amplifier noise secondary. In still other cases, the noise combination of the amplifier noise and the feedback resistance dominates noise performance through noise gain peaking. The multidimensional aspects of this noise performance precludes fixed guidelines that would replace numerical evaluation and comparison.

To illustrate this point, consider varying just the R_f variable while examining the E_{no} result. This displays a noise dominance contest between the amplifier input noise voltage, the resistor noise, and noise gain peaking. For simplicity, this illustration case ignores the amplifier's input noise current, as often permitted by the low noise current of an FET-input op amp. This simplification reduces the net output noise to $E_{no} = \sqrt{E_{noR}^2 + E_{noe}^2}$. For this case, Fig. 5.7 shows the current-to-voltage converter's dominant noise effects for three characteristic ranges of feedback resistance. The curve there represents total output noise for the basic current-to-voltage converter of Fig. 5.1 under the influence of the characteristic noise gain of Fig. 5.3.

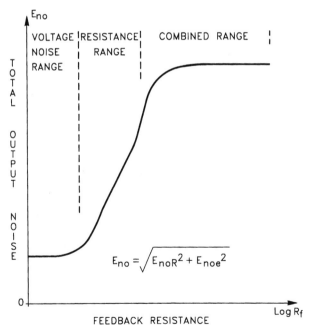

Figure 5.7 As the feedback resistance increases, the dominant noise source of a current-to-voltage converter changes from the op amp to the resistor and then to the combination.

Different factors dominate the E_{no} curve for different ranges of feedback resistance. At low resistance levels, the curve remains flat with the op amp voltage noise the dominant contributor, producing a voltage-noise range of dominance. There, noise gain peaking has not yet developed, and resistor noise remains in the background, so E_{no} remains small. This region typically covers the feedback resistance range of 0 to 10 kΩ. In the next region, resistor noise overtakes the voltage component, producing a rising curve and a resistance range of dominance. There, the curve's rising slope initially displays the square root noise-versus-resistance relationship of a resistor. This region typically covers the feedback resistance range of 10 kΩ to 1 MΩ.

Even higher resistances introduce noise gain peaking, increasing the upward slope of the E_{no} curve at the beginning of a range of combined influence. There, the voltage noise of the amplifier, amplified by gain peaking, overrides the resistor noise to increase this slope. Eventually, stray capacitance bypass of the feedback resistance terminates this rise, producing a level ending for the E_{no} curve. In this leveled region, the resistor and voltage noise effects, E_{noR} and E_{noe} of E_{no}, reduce to constants. E_{noR} becomes a constant because any further increase in the R_f resistance produces a matching decline in the associated bandwidth of the R_f, C_S combination. There, a resistance increase raises the resistor's noise density, $e_{nR} = \sqrt{4KTR_f}$ but reduces the associated noise bandwidth $BW_t = 1/2\pi R_f C_S$. The two changes produce canceling effects in the rms output $E_{noR} = \sqrt{2KTR_f \pi BW_t}$, holding E_{noR} constant.

E_{noe} becomes a constant because the presence of C_S limits the noise gain A_{ne} to a plateau level in the previous Fig. 5.3. There, the stray capacitance limits the gain peaking to a final value of $1 + C_i/C_S$, regardless of any further increase in the value of R_f. This makes $E_{noe} = A_{ne}e_{nif} = (1 + C_i/C_S)e_{nif}$ and holds E_{noe} constant at higher frequencies. Still, the continued increase of R_f does expand the frequency span of the peak A_{ne} level in Fig. 5.3. There, increasing R_f moves the $f_{zf} = 1/2\pi R_f(C_i + C_S)$ starting point of the gain peaking to the left, expanding the span of A_{ne} peaking. However, this expansion produces diminishing returns as revealed by two numerical examples. For a peak A_{ne} level starting at 200 kHz, doubling R_f reduces the starting point to 100 kHz, adding 100 kHz of span to the peak A_{ne} level. Similarly, for a peak level starting at 2 kHz, doubling R_f reduces the starting point to 1 kHz, adding 1 kHz of span. Thus, the same doubling action produces only 1/100th the span increase in the second example. The diminishing increase in noise bandwidth results in a nearly constant E_{noe} versus R_f response as R_f continues to increase in the final stage of Fig. 5.7. Together, the constant E_{noR} and E_{noe} responses level the E_{no} versus R_f response shown there.

Reflection upon this response reveals two cases where choosing an op amp for low input noise may or may not optimize noise performance for the photodiode amplifier. First, in the resistance range, the noise of the feedback resistor dominates, making the amplifier noise voltage effect insignificant. Second, in the combined range, the amplifier noise voltage returns to significance, but the amplifier's input capacitance often plays an equally significant role in noise performance. There, noise gain peaking controls the gain applied to the noise voltage. This peaking results from the capacitance at the circuit input as well as from the high feedback resistance. At the circuit input, amplifier input capacitance adds to that of the photodiode, increasing the gain-peaking effect. An inherent design compromise in FET op amp design tends to increase input capacitance as it decreases noise.

For photodiode applications, FET-input op amps most commonly serve the current-to-voltage converter's requirement for low amplifier input current. With these amplifiers, decreasing input noise requires increasing the size of the amplifier's input FETs to raise their transconductances. Transconductance increases in direct proportion to the FET size but so does the FET capacitance. The greater FET capacitance increases the gain peaking effect in both magnitude and bandwidth. This compromise becomes very significant when using lower-capacitance photodiodes. In such cases, a lower-noise amplifier can produce greater noise gain peaking, sometimes degrading rather than improving noise performance.

Some degree of trial and error characterizes the noise design of a photodiode amplifier due to the numerous variables involved. It would take a five-dimensional graph to display the direct and indirect noise effects of the resistor combined with those of the amplifier and the photodiode. The noise performance of a given application must be numerically evaluated to define the significance of each noise factor. Otherwise, a design focused upon one factor may suboptimize overall performance, as described in the preceding examples. For a specific application, starting with an initial design and iterating from there reduces the ranges of the numerous variables and eases the design optimization. For each iteration, numerical evaluation of E_{no} and comparison of its component terms identifies the opportunities for improving the performance of a given case. Where amplifier noise voltage remains a significant factor, a variety of circuit options minimize its predominant or compromising effect as described in later chapters.

References

1. G. Tobey, J. Graeme, and L. Huelsman, *Operational Amplifiers; Design and Applications*, McGraw-Hill, New York, 1971.
2. S. Millaway and D. Haynes, OPA101 Product Data Sheet, PDS434A, Burr-Brown Corp., 1980.
3. J. Graeme, "Divide and Conquer Noise in Photodiode Amplifiers," *Electronic Design*, June 27, 1994, p. 10.
4. D. A. Bell, *Noise and the Solid State*, John Wiley & Sons, New York, 1985.

Chapter

6

Noise Reduction

Higher-gain photodiode amplifiers typically employ very large feedback resistances that make the noise gain peaking of Chap. 5 the circuit's dominant noise effect. For a given application, performing the noise analysis of Sec. 5.3.1 determines if this is the case through the relative magnitudes of the E_{noe}, E_{noR}, and E_{noi} noise components. When the E_{noe} result exceeds the other two, gain peaking dominates noise performance. As will be shown, the circuit then amplifies the noise voltage of the op amp's input with a noise bandwidth that exceeds the signal bandwidth. Three modifications to the basic photodiode amplifier provide filtering to reduce or remove this noise disadvantage.[1] In each case, noise analysis equations guide the filter component selection to optimize the circuit's noise-versus-bandwidth compromise.

The three circuit alternatives reduce the E_{noe} noise component with differing circuit complexities and signal bandwidth results. The first simply adds feedback capacitance, reducing the noise gain peak but also reducing signal bandwidth. The other two alternatives apply greater circuit complexity to better address the noise bandwidth issue. They reduce noise bandwidth without greatly affecting the signal bandwidth. The second filtering alternative adds an op amp and feedback elements to form a noise-filtering composite amplifier. This composite alternative removes the input error signals otherwise introduced by the second op amp of a traditional filter. The third alternative avoids the added amplifier by exploiting the filtering effects of a common phase-compensation method. However, this simplified alternative requires empirical determination of the circuit's phase compensation.

6.1 Noise Reduction with Feedback Capacitance C_f

In the first approach to noise reduction, simply adding feedback capacitance to the photodiode amplifier reduces the circuit's high-frequency noise gain. This does not actually reduce noise bandwidth, but the reduced gain peaking decreases the high-frequency effect of the op amp's noise voltage. Also, the added C_f capacitor appears in parallel with the stray C_S considered in earlier noise analyses, permitting an easy adjustment of previous results for this case. However, adding C_f also reduces signal bandwidth. Also, this approach reaches a point of diminishing returns because continued increase of C_f can only reduce the noise gain to unity.

6.1.1 The noise gain reduction of C_f

Figure 6.1 models this case with the equivalent circuit of a photodiode connected to a current-to-voltage converter. This figure models the photodiode with a current source i_p and a shunt capacitance C_D included in C_i. Capacitance C_i also includes the input capacitances C_{id} and C_{icm} of the op amp, making the net capacitance of the input

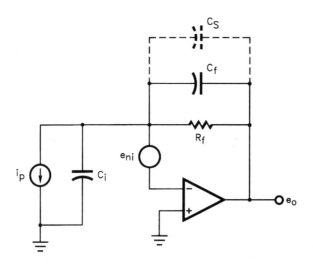

$$e_{noe} = \frac{1 + R_f C_i s}{1 + R_f(C_S + C_f)s} e_{ni} \qquad C_i = C_D + C_{id} + C_{icm}$$

Figure 6.1 Adding feedback capacitance C_f increases the capacitive shunting of R_f to reduce the high-frequency noise gain presented to amplifier noise voltage e_{ni}.

circuit $C_i = C_D + C_{id} + C_{icm}$. Op amp noise voltage e_{ni} appears across C_i, producing a noise current that feedback forces through feedback resistor R_f. The resulting output noise voltage e_{noe} underlies the noise gain peaking of the common photodiode amplifier. As described in Chap. 5, rms integration of e_{noe} over its noise bandwidth produces the E_{noe} result of Sec. 5.3.1. For the high-gain case, stray capacitance C_S automatically bypasses R_f to limit this noise gain peaking at a level of $1 + C_i/C_S$. For greater noise reduction, adding capacitor C_f further restricts this gain peak to a level of $1 + C_i/(C_f + C_S)$.

Figure 6.2 shows the resulting response curves superimposed upon the original noise analysis plot of Fig. 5.3. Once again, this figure displays the amplifier's open-loop gain A_{OL}, the I-to-V gain, and the noise gain A_{ne} that amplifies the op amp's input noise voltage. The bold curves and the primed notations represent the modifications introduced by C_f. Adding C_f changes four characteristics of the noise gain response A_{ne}, converting it to the A_{ne}' response shown. However, the general shape of the A_{ne}' response remains unchanged, permitting extrapolation of the Chap. 5 noise results to this case. As desired, C_f lowers the peak level of the noise gain, removing the shaded area of the figure from the noise response. In doing so, C_f alters three of the

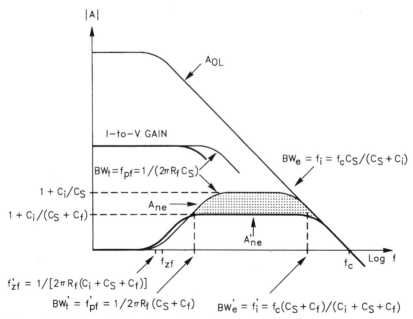

Figure 6.2 Noise reduction through the C_f bypass of Fig. 6.1 removes the shaded area of noise gain but also reduces signal bandwidth from BW_t to BW_t'.

curve's characteristic frequencies, f_{pf}, f_i, and f_{zf}. The first two frequencies define the circuit's signal and noise bandwidths, originally $BW_t = f_{pf}$ and $BW_e = f_i$. BW_t represents the response limit of the circuit's transresistance gain, and BW_e represents this limit for A_{ne}. Adding C_f converts these limits to $BW_t' = f_{pf}'$ and $BW_e' = f_i'$, reducing the first limit and increasing the second as shown. The BW_t to BW_t' conversion reduces signal bandwidth, making the bold I-to-V gain curve roll off earlier. The BW_e to BW_e' conversion increases noise bandwidth, but this actually decreases rather than increases noise. Comparison of the A_{ne} and A_{ne}' curves shows that expanding the shaded area removed from the noise response simply moves BW_e to a higher frequency.

The third frequency modified by the gain peak reduction f_{zf} does reflect a noise increase but one of little consequence. This frequency marks the beginning of the A_{ne} noise peak, and the addition of C_f moves this frequency from $f_{zf} = 1/2\pi R_f(C_i + C_S)$ to $f_{zf}' = 1/2\pi R_f(C_i + C_S + C_f)$. The A_{ne}' curve reflects this change through its earlier rise, and this increases the area under the noise gain curve. However, two factors limit the effect of this increase. The relative increase generally remains small and it occurs in the low-frequency portion of the gain peak. Generally, $C_f \ll C_i + C_S$, making the shift in f_{zf} a minor change. Further, the noise response increase occurs in a region of reduced significance, as displayed by the earlier comparison of Figs. 5.5 and 5.6. There, conversion from logarithmic to linear frequency scales showed that lower-frequency regions of the noise response add relatively little to the net circuit noise. Numerical analysis of the specific design case reveals the net noise reduction of the C_f bypass.

6.1.2 Noise analysis for the C_f case

One substitution, in multiple places, adapts the noise analysis summary of Sec. 5.3.1 to this added capacitance case. Because C_f connects in parallel with C_S, simply substituting $C_f + C_S$ for C_S includes the added effect of C_f in the earlier analysis results. A modified summary follows, reflecting this substitution for noise analysis of the Fig. 6.1 circuit. This substitution changes E_{noR}, E_{noi}, and all of the E_{noe} components except E_{noe1}. E_{noR} and E_{noi} change due to their new bandwidth limit $BW_t' = 1/2\pi R_f(C_S + C_f)$. E_{noe2} through E_{noe5} change due to the new region-defining frequencies $f_{zf}' = 1/2\pi R_f(C_i + C_S + C_f)$, $f_{pf}' = 1/2\pi R_f(C_S + C_f)$ and $f_i' = f_c(C_S + C_f)/(C_i + C_S + C_f)$. E_{noe4} also changes due to the new peak gain level of $1 + C_i/(C_f + C_S)$. Otherwise, the noise analysis results remain the same in spite of the mathematical integration underlying the previous rms analyses. The integration results still hold because the addition of C_f does not change the characteristic shapes of the A_{ne} and I-to-V gain curves in Fig. 6.2. As a result, just modifying the frequencies that define various integration limits and adjusting the expression for the peak gain level adapts the previous results to the C_f case.

Making these changes to the noise analysis summary of Sec. 5.3.1 produces the analysis procedure below for Fig. 6.1. For a specific application, step-by-step numerical evaluation of the noise components below defines the relative significance of each and concludes with a total rms output noise result.

$E_{\text{noe1}} = e_{\text{nif}}\sqrt{f_f \ln(f_f/f_1)}$, where e_{nif} is the e_{ni} noise floor level, and f_f is the $1/f$ noise corner frequency. As explained in Sec. 5.2.3, setting $f_1 \approx 0.01$ Hz avoids a theoretical contradiction and ensures analytical accuracy.

$E_{\text{noe2}} = e_{\text{nif}}\sqrt{f_{\text{zf}}' - f_f}$, where $f_{\text{zf}}' = 1/2\pi R_f(C_i + C_S + C_f)$, and $C_i = C_D + C_{\text{id}} + C_{\text{icm}}$. Here and in the E_{noe4} analysis, numerical evaluation can produce the square root of a negative number. Then, substitute zero for the result as explained in Sec. 5.3.1.

$E_{\text{noe3}} = (e_{\text{nif}}/f_{\text{zf}}')\sqrt{(f_{\text{pf}}'^3 - f_{\text{zf}}'^3)/3}$, where $f_{\text{pf}}' = 1/2\pi R_f(C_S + C_f)$.

$E_{\text{noe4}} = [1 + C_i/(C_S + C_f)]e_{\text{nif}}\sqrt{f_i' - f_{\text{pf}}'}$, where $f_i' = f_c(C_S + C_f)/(C_i + C_S + C_f)$.

$E_{\text{noe5}} = (e_{\text{nif}}f_c)\sqrt{1/f_i'}$, where f_c is the op amp's unity-gain crossover frequency.

$E_{\text{noe}} = \sqrt{E_{\text{noe1}}^2 + E_{\text{noe2}}^2 + E_{\text{noe3}}^2 + E_{\text{noe4}}^2 + E_{\text{noe5}}^2}$, the total rms output noise resulting from the spectral noise voltage e_{ni} of the amplifier input.

$E_{\text{noR}} = \sqrt{2KTR_f \pi \text{BW}_t'}$, where $\text{BW}_t' = 1/2\pi R_f(C_S + C_f)$.

$E_{\text{noi}} = R_f\sqrt{q\pi \text{BW}_t' I_{B-}}$

$E_{\text{no}} = \sqrt{E_{\text{noR}}^2 + E_{\text{noi}}^2 + E_{\text{noe}}^2}$.

The BW_t' modification above reveals the fundamental drawback of the added bypass of C_f. While this capacitance reduces the noise gain peak, it also shunts the circuit's signal current, reducing signal bandwidth. As long as bandwidth requirements permit, this noise reduction alternative offers the simplest solution in terms of added circuit elements. However, this solution does not necessarily provide the greatest reduction in total output noise. Addition of C_f reduces the peak level of A_{ne} but does not effectively restrict noise bandwidth. Like A_{ne}, the modified A_{ne}' response continues until rolled off by the amplifier's A_{OL} response. Restricting noise bandwidth instead provides greater noise reduction through greater circuit complexity as described below.

6.2 Noise Bandwidth versus Signal Bandwidth

A more effective noise reduction results from restricting noise bandwidth without significantly reducing signal bandwidth. Unlike most op amp circuit's, the high-gain photodiode amplifier produces a bandwidth distinction for signal and noise. This distinction permits selective reduction of noise without affecting signal. As shown in Fig. 6.2, one bandwidth limit, BW_t, governs the roll off of the circuit's transresistance or I-to-V gain. Another limit, BW_e, governs the roll of the noise gain A_{ne} that amplifies input noise voltage e_{ni}. Comparison of the I-to-V and A_{ne} curves reveals a noise gain continuation long after the signal gain rolls off. There, noise gain A_{ne} continues until interrupted by the op amp's A_{OL} roll off. As a result, the majority of the op amp's bandwidth often serves to amplify noise and not the signal. Then, reducing the op amp's open-loop bandwidth decreases noise without affecting signal. Ideally, reducing the amplifier bandwidth to the same frequency as the signal bandwidth limit optimizes noise performance. Op amps with provision for external phase compensation automatically offer this bandwidth-limiting option. However, the limited availability of this provision greatly restricts amplifier choices. The two remaining filtering alternatives below remove this restriction and produce the ideal, coincident signal and noise bandwidths limits. One uses a filtering composite amplifier, and the other uses a decoupling phase compensation.

6.3 Noise Reduction with a Composite Amplifier

The second filtering alternative adds an op amp and feedback elements to form a noise-filtering composite amplifier. Adding a modified integrator within the feedback loop of the photodiode amplifier permits the roll off of the higher-frequency open-loop gain that otherwise only supports noise bandwidth. Design equations guide the design of this composite amplifier to optimize noise reduction while preserving frequency stability. Once again, extrapolation of earlier noise results permits simplified noise performance analysis here.

6.3.1 The noise bandwidth reduction of the composite amplifier

A composite amplifier configuration equalizes the noise and signal bandwidths by adding a second op amp for noise bandwidth control.

Shown in Fig. 6.3, this configuration still permits selection of amplifier A_1 for low input current and low noise without regard for an external phase-compensation provision. Amplifier A_2 and its local feedback substitute for this provision by modifying the composite circuit's open-loop response. Adding A_2 and its local feedback produces higher-frequency attenuation, removing the gain that otherwise only supports noise bandwidth. Note that the two op amps in series here requires return of the R_f feedback to A_1's noninverting rather than inverting input. This modified return avoids enclosing the two amplifiers' phase inversions in the same feedback loop, retaining negative feedback.

Here, A_2 varies from an amplifier to an integrator and then to an attenuator as frequency varies. At low frequencies, C_1 blocks A_2's local feedback, and this amplifier contributes its full open-loop gain to the composite feedback. There, the increased gain reduces low-frequency error. At intermediate frequencies, the integrator feedback formed by R_1 and C_1 reduces the A_2 gain support in a transition to the attenuator mode. At high frequencies, C_1 becomes a short circuit, relinquishing A_2's gain control to the closed-loop effects of an R_1, R_2 feedback. There, $A_{CL2} = -R_2/R_1$, and making $R_2 < R_1$ produces the

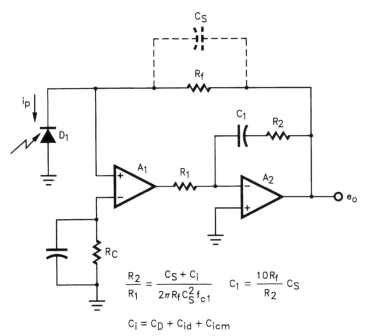

$$\frac{R_2}{R_1} = \frac{C_S + C_i}{2\pi R_f C_S^2 f_{c1}} \qquad C_1 = \frac{10 R_f}{R_2} C_S$$

$$C_i = C_D + C_{id} + C_{icm}$$

Figure 6.3 Adding a modified integrator within the feedback loop of the photodiode amplifier permits roll off of the higher-frequency open-loop gain that otherwise only supports noise bandwidth.

desired high-frequency attenuation. The composite amplifier structure leaves other characteristics of the current-to-voltage converter virtually unchanged. Adding A_2 introduces no additional noise or offset because the high gain of A_1 isolates the composite circuit input from A_2's input errors.

Figure 6.4 shows the effect of the composite amplifier upon the circuit's net open-loop gain and bandwidth. There, the A_{OL1} curve of A_1 follows the normal open-loop response, and the A_{CL2} curve of A_2 follows a modified integrator response. As described previously, A_{CL2} produces an integrator characteristic at intermediate frequencies until interrupted by a response zero at $f_{z2} = 1/2\pi R_2 C_1$. There, the inclusion of R_2 in the A_2 local feedback converts the A_{CL2} curve to the flat response of a voltage amplifier. That voltage amplifier produces a gain magnitude of R_2/R_1, and making $R_2/R_1 < 1$ places this magnitude below the unity-gain or 0-dB axis. With a gain magnitude less than unity, A_{CL2} produces signal attenuation in the higher-frequency range of the A_{CL2} response. This attenuation transfers to the circuit's composite response due to the series connection of the two amplifiers.

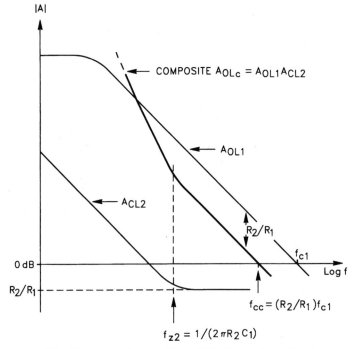

Figure 6.4 The composite amplifier of Fig. 6.3 replaces the A_{OL1} response of the basic photodiode amplifier with A_{OLc}, removing high-frequency open-loop gain.

Combined, the two amplifiers produce the $A_{OLc} = A_{OL1}A_{CL2}$ response shown in bold.

The composite's attenuation drops the A_{OLc} response below the A_{OL1} curve of a single amplifier solution, reducing the high-frequency open-loop gain. This gain reduction equals the R_2/R_1 attenuation factor of A_{CL2} and produces an equal reduction in high-frequency bandwidth. This equality results from the single-pole roll offs of A_{OLc} and A_{OL1} following f_{z2}. Single-pole roll offs follow a 1:1 slope, so a given vertical move of the response curve produces an equal horizontal move. Thus, the R_2/R_1 drop in gain magnitude produces an equal decrease in unity-gain crossover frequency, moving the crossover from f_{c1} to the composite $f_{cc} = (R_2/R_1)f_{c1}$. Together, the gain and bandwidth reductions move the circuit's open-loop roll off down in frequency, much as if the original op amp bandwidth had been reduced. Thus, the composite amplifier replaces the external phase compensation option of the single-amplifier solution. In addition, the composite amplifier boosts low-frequency open-loop gain for increased response accuracy there.

6.3.2 The noise gain reduction of the composite amplifier

Figure 6.5 demonstrates the composite's noise reduction by comparing its response with that of the basic current-to-voltage converter. There, the bold curves and c subscripts denote the modifications produced by the composite. This figure combines the composite amplifier response of Fig. 6.4 with the photodiode amplifier's basic response curves from Fig. 5.3. As desired, the reduced gain-bandwidth product of A_{OLc} rolls off noise gain A_{ne} at a reduced intercept frequency f_{ic}. Note that the new A_{ne} roll off also drives this gain below the unity-gain or 0-dB axis. This avoids the continued $A_{ne} = 1$ minimum encountered with the previous C_f alternative. Also, the signal or I-to-V response here retains its original bandwidth limit at the frequency f_{pf}. As long as $f_{ic} \geq f_{pf}$, the composite amplifier reduces noise bandwidth without significantly affecting signal bandwidth.

The composite's A_{ne} modification eliminates the shaded area of noise gain, which the logarithmic frequency scale visually compresses. Actually, the resulting noise reduction greatly exceeds the visual perception because most of the amplifier's bandwidth lies compressed in this upper end of the logarithmic response curve. For better appreciation of the relative noise reduction, comparison of the earlier Figs. 5.5 and 5.6 reveals the dominance of high-frequency effects upon the noise result. For another comparison, consider moving the roll-off frequency of the noise gain f_i back a decade in frequency to the f_{ic} shown here. This move typically extends the shaded area of the figure over a minor

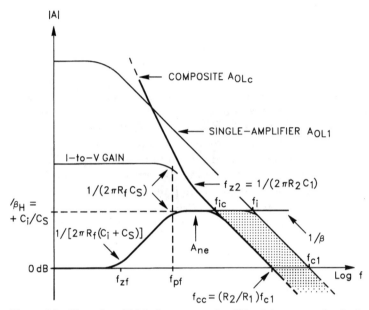

Figure 6.5 The reduced high-frequency gain of Fig. 6.4 removes the shaded area from the A_{ne} noise response without altering the I-to-V bandwidth limit at f_{pf}.

portion of the A_{ne} response shown. However, this decade move reduces the output noise produced by A_1 by about a factor of 3. In many cases, a reduction of this magnitude eliminates the significance of the amplifier noise.

Removing the shaded area under the A_{ne} curve leaves the general shape of the curve unchanged, and the E_{noe} results of the basic photodiode amplifier still apply. However, two frequency variables expressed in those results change for this case. In the figure, f_{cc} and f_{ic} define A_{OLc} intercepts that were previously marked by f_c and f_i with A_{OL1}. Defining these two new frequencies in terms of the known, crossover frequency f_{c1} translates the Sec. 5.3.1 noise results to this composite case. As described with Fig. 6.4, $f_{cc} = (R_2/R_1)f_{c1}$ due to the R_2/R_1 attenuation provided by A_2 at higher frequencies. This attenuation factor also relates f_{ic} to f_i due to the Fig. 6.5 parallelogram marked by f_{ic} and f_i in conjunction with f_{cc} and f_{c1}. There, frequencies f_{ic} and f_i mark one side of this parallelogram, opposite the side marked by f_{cc} and f_{c1}. The parallelogram makes two sides have equal length and, accounting for the logarithmic scaling of response plots, $\text{Log}(f_i) - \text{Log}(f_{ic}) = \text{Log}(f_{c1}) - \text{Log}(f_{cc})$. Then, $f_i/f_{ic} = f_{c1}/f_{cc}$, and solving for f_{ic} produces $f_{ic} = (f_{cc}/f_{c1})f_i = (R_2/R_1)f_i$. Here, f_i represents the intercept frequency of the original A_{OL1} response and the circuit's $1/\beta$ curve. The typical single-pole A_{OL1} roll off makes $f_i = \beta_H f_{c1}$ for $f_{ic} = (R_2/R_1)\beta_H f_{c1}$. Here, β_H is the high-frequency value for the feedback factor β and from before, $\beta_H = C_S/(C_i + C_S)$. Thus, $f_{ic} = f_{cc}/(1 + C_i/C_S)$, defining f_{ic}.

6.3.3 Optimizing the composite noise-versus-bandwidth compromise

The f_{ic} and f_{cc} expressions developed above define the modification of the A_{ne} response produced by the composite amplifier. Both expressions define response frequencies in terms of specific circuit elements and parameters. However, the R_2/R_1 factor of the two expressions remains arbitrary at this point. In practice, this arbitrary factor presents an extra degree of design freedom. Optimum selection of the R_2/R_1 factor produces a compromise between noise and signal bandwidths through placement of the modified intercept frequency f_{ic}. In Fig. 6.5, decreasing R_2/R_1 moves f_{ic} to the left, expanding the shaded area removed from the noise response. However, this move also brings a new signal bandwidth limit into significance. Frequency f_{ic} represents the crossing of the gain demand curve of $1/\beta$ and the available gain curve of A_{OLc}. Both f_{ic} and f_{pf} impose bandwidth limits with the lower of the two controlling the actual result. In compromise, setting $f_{ic} = f_{pf}$ makes the two limits coincident for the maximum possible noise reduction without a major reduction of signal bandwidth. For this compromise, the two coincident poles at f_{pf} reduce the -3-dB signal bandwidth from f_{pf} to $0.64 f_{pf}$.

The $f_{ic} = f_{pf}$ compromise and a phase compensation requirement define the design equations for the composite amplifier. First, analysis of the compromise defines the R_2/R_1 attenuation ratio. Then, choosing a convenient R_2 value determines the R_1 value as well. Finally, an examination of phase compensation requirements defines the design equation for C_1, also a function of the initial R_2 choice. Equating $f_{ic} = f_{pf}$ defines R_2/R_1 through prior equations for the two frequencies. From Chap. 5, $f_{pf} = 1/2\pi R_f C_S$ represents the basic feedback pole produced by stray capacitance bypass of R_f. Frequency f_{ic} represents the shifted $1/\beta$ intercept of the composite circuit as shown in Fig. 6.5. From Sec. 6.3.2, $f_{ic} = f_{cc}/(1 + C_i/C_S) = \beta_H f_{cc}$, where $f_{cc} = (R_2/R_1)f_{c1}$, making $f_{ic} = \beta_H (R_2/R_1) f_{c1}$. Equating the f_{ic} and f_{pf} expressions and solving for the R_2/R_1 attenuation ratio yields

$$\frac{R_2}{R_1} = \frac{f_{pf}}{\beta_H f_{c1}}$$

where $f_{pf} = 1/2\pi R_f C_S$, $\beta_H = C_S/(C_S + C_i)$, and f_{c1} is the unity-gain crossover frequency of A_1. Then choosing R_2 for a convenient resistance level defines R_1 through

$$R_1 = \frac{\beta_H f_{c1}}{f_{pf}} R_2$$

Next, the circuit's phase compensation requirement defines the C_1 value required to stabilize the circuit. The composite amplifier's inherent two-pole A_{OLc} response corresponds to 180° of feedback phase shift and potentially produces oscillation. Phase compensation preserves stability by reducing this slope where the A_{OLc} curve intercepts the circuit's $1/\beta$ response in Fig. 6.5. Two- to single-pole slope conversion results from the A_{OLc} zero, at f_{z2}. Selection of C_1 defines this zero at the break frequency of C_1 with R_2, producing $f_{z2} = 1/2\pi R_2 C_1$. Initiating this slope conversion well before the $1/\beta$ intercept at f_{ic} avoids any significant stability degradation. This reduces the A_{OLc} phase shift at f_{ic} to that normally encountered with single-amplifier circuits. In practice, making $f_{z2} = f_{ic}/10$ virtually removes the added phase shift of the composite amplifier. Combining this frequency relationship with the earlier $f_{ic} = f_{pf}$ compromise produces $f_{z2} = f_{pf}/10$, where $f_{z2} = 1/2\pi R_2 C_1$ and $f_{pf} = 1/2\pi R_f C_S$. Finally, combining these last three relationships and solving for C_1 yields the design equation

$$C_1 = \frac{10 R_f}{R_2} C_S$$

6.3.4 Noise analysis for the composite case

The $f_{ic} = f_{pf}$ choice for the composite photodiode amplifier reduces four of the seven noise components of the basic photodiode amplifier. The components E_{noe4}, E_{noe5}, E_{noR}, and E_{noi} of the Sec. 5.3.1 summary all decrease due to modified frequency roll offs for their underlying noise sources. The composite alternative reduces the roll off frequency for the E_{noe4} and E_{noe5} components and adds a second pole to the roll offs of E_{noR} and E_{noi}.

E_{noe4} and E_{noe5} decrease due to the reduced frequency response of noise gain A_{ne} in Fig. 6.5. This gain amplifies the op amp's input noise voltage as analyzed in five frequency regions in Sec. 5.2.3. The composite moves the A_{ne} roll off back in frequency, reducing the previous noise contributions of regions 4 and 5 to E_{noe}. For region 4, the previous analysis defined the noise component $E_{noe4} = (1 + C_i/C_S)e_{nif}\sqrt{f_{ic} - f_{pf}}$. Here, making $f_{ic} = f_{pf}$ places the starting and ending points of region 4 at the same frequency, eliminating the region, and $E_{noe4} = 0$. This is the result of making the noise and signal bandwidths coincide. For region 5, reducing the high-frequency A_{ne} gain moves the reference frequencies f_{c1} and f_i to the composite's f_{cc} and f_{ic} of Fig. 6.5. This makes $E_{noe5} = (e_{nif}f_{cc})\sqrt{1/f_{ic}}$, where f_{cc} represents the reduced unity-gain crossover frequency of the composite as described in Sec. 6.3.2. There, the circuit's attenuation factor R_2/R_1 reduces this crossover to $f_{cc} = (R_2/R_1)f_{c1}$. Making $f_{ic} = f_{pf}$ for the above noise-versus-bandwidth compromise simplifies the result by removing the f_{ic} variable. Then, $E_{noe5} = (e_{nif}f_{cc})\sqrt{1/f_{pf}}$, where $f_{pf} = 1/2\pi R_f C_S$.

The reduction of the other two noise components, E_{noR} and E_{noi}, results from the two-pole roll off now presented to the feedback resistor noise and the amplifier input current noise. Previously, a single pole at f_{pf} rolled off these noise components through stray capacitance bypass of the feedback resistor. The composite's $f_{ic} = f_{pf}$ compromise adds the roll off of the $1/\beta$ intercept at f_{ic}, producing a two-pole roll off for these noise components. The added pole changes the noise response shape, requiring reevaluation of the Sec. 5.1.3 rms integral

$$E_{no}^2 = \int_0^\infty |A_n e_n|^2 \, df$$

This integral expresses the rms output noise E_{no} resulting from a noise source e_n in the presence of a noise gain A_n. The feedback resistor produces the noise source $e_{nR} = \sqrt{4KTR_f}$ by itself and $e_{nRi} = R_f\sqrt{2qI_{B-}}$ in conjunction with the amplifier's input noise current. As described before, the photodiode amplifier transfers these noise components to the circuit output with unity gain at lower frequencies. At higher frequencies, the composite photodiode amplifier rolls off these noise effects with a double pole at $f_{ic} = f_{pf}$. Together, the initial unity gain and the two-pole roll off make $A_n = 1/(1 + jf/f_{pf})^2$ for the E_{noR} and E_{noi} rms analyses. Performing the two rms integrations produces

$$E_{noR} = \sqrt{KTR_f \pi f_{pf}} \quad \text{and} \quad E_{noi} = R_f\sqrt{q \pi f_{pf} I_{B-}/2}$$

These results indicate a $\sqrt{2}$ reduction in E_{noR} and E_{noi} as compared with the basic photodiode amplifier of Sec. 5.1.3.

Modifying the Sec. 5.3.1 summary with the four new results above produces the listing below for step-by-step noise analysis of the composite photodiode amplifier of Fig. 6.3.

$E_{noe1} = e_{nif}\sqrt{f_f \ln(f_f/f_1)}$, where e_{nif} is the e_{ni} noise floor level, and f_f is the $1/f$ noise corner frequency. As explained in Sec. 5.2.3, setting $f_1 \approx 0.01$ Hz avoids a theoretical contradiction and ensures analytical accuracy.

$E_{noe2} = e_{nif}\sqrt{f_{zf} - f_f}$, where $f_{zf} = 1/2\pi R_f(C_i + C_S)$, and $C_i = C_D + C_{id} + C_{icm}$. Here and in the E_{noe4} analysis, numerical evaluation can produce the square root of a negative number. Then, substitute zero for the result as explained in Sec. 5.3.1.

$E_{noe3} = (e_{nif}/f_{zf})\sqrt{\left(f_{pf}^3 - f_{zf}^3\right)/3}$, where $f_{pf} = 1/2\pi R_f C_S$.

$E_{noe4} = 0$.

$E_{noe5} = (e_{nif}f_{cc})\sqrt{1/f_{pf}}$, where $f_{cc} = (R_2/R_1)f_{c1}$, and f_{c1} is the unity-gain crossover frequency of A_1.

$E_{noe} = \sqrt{E_{noe1}^2 + E_{noe2}^2 + E_{noe3}^2 + E_{noe5}^2}$, the total rms output noise resulting from the spectral noise voltage e_{ni} of the amplifier input.

$E_{noR} = \sqrt{KTR_f \pi f_{pf}}$.

$E_{noi} = R_f\sqrt{q\pi f_{pf} I_{B-}/2}$.

$E_{no} = \sqrt{E_{noR}^2 + E_{noi}^2 + E_{noe}^2}$.

6.3.5 A comparison with the active filter alternative

The noise reduction of the composite amplifier above decreases amplifier bandwidth to remove that portion that serves noise but not signal. The more obvious approach, adding an active filter after the photodiode amplifier, produces the same or even an improved noise result but sacrifices an error reduction feature of the composite solution. A simplified filtering approach later retains the composite's error reduction and removes one amplifier from the circuit in Sec. 6.4. However, this simplified circuit may either increase or decrease the noise reduction, depending upon the specific application.

As an obvious alternative to the composite solution, placing an active filter following the basic current-to-voltage converter permits a wide variety of bandwidth-shaping options. Then, setting the filter response to the frequency range of useful information again eliminates that bandwidth that only serves noise. Like the composite solution, this active filter alternative only requires the addition of a second op amp and its feedback components. Further, this alternative permits the addition of multiple poles for steeper roll off of noise bandwidth. However, unlike the composite solution, this filter alternative adds the input error effects of the added amplifier directly in the signal path. Such a filter resides outside the current-to-voltage converter's feedback loop, removing one of the composite amplifier's primary benefits. The common feedback loop of the composite's two amplifiers removes

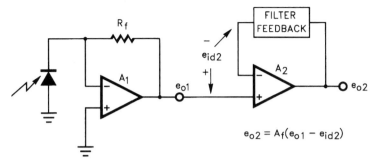

Figure 6.6 Simply adding a filter amplifier after the photodiode amplifier again removes higher-frequency noise with an added op amp but sacrifices the composite amplifier's attenuation of the e_{id2} error components.

the input error effects of the second amplifier through the gain isolation provided by the first. There, the offset, noise, and multiple other input-referred errors of the second amplifier reflect to the composite input divided by the high, open-loop gain of the first amplifier. The active filter alternative operates with an independent feedback loop, sacrificing this error reduction feature as illustrated by Fig. 6.6. There, an arbitrary active filter formed with A_2 follows the current-to-voltage converter formed with A_1. This configuration adds the e_{id2} error signal of A_2 directly in the signal path. This error signal represents the combined input errors of A_2 due to offset voltage and noise plus the error signals resulting from finite open-loop gain and power-supply rejection ratio (PSRR). From the figure, $e_{o2} = A_f(e_{o1} - e_{id2})$ reflects the e_{id2} error influence upon the circuit's final output signal where A_f represents the gain of the active filter.

6.4 Noise Reduction with Decoupling Phase Compensation

The third filtering alternative for photodiode amplifiers avoids the added amplifier of the composite amplifier and active filter by exploiting the filtering effects of a common phase-compensation method. Decoupling phase compensation reduces the noise gain of a photodiode amplifier by adding a passive filter within the feedback loop of a single-amplifier current-to-voltage converter. This decoupling approach does not reduce the noise bandwidth as much as the composite alternative, but the decoupling rolls off the noise gain peaking with a two-pole rather than a single-pole slope. As a result, this approach may either degrade or improve the noise reduction. For a specific application, comparative noise analyses of the two techniques reveals the relative benefits of the two. However, where the single-amplifier approach prevails, this simplified alternative requires empirical determination of the circuit's phase compensation.

6.4.1 The noise gain reduction of the decoupling

Shown in Fig. 6.7, this noise filtering circuit originates in the phase compensation technique commonly used to accommodate capacitance loads.[2] Applied to the photodiode case, this phase compensation produces a filtering photodiode amplifier. In the figure, an intentionally added capacitance load C_L combines with resistor R_C and amplifier output resistance R_o to form a low-pass filter. This filter resides between the outputs of the amplifier and the overall circuit, and shunts high-frequency noise to ground. This shunting to ground produces a much greater noise attenuation than the simple feedback bypass of Fig. 6.1. There, bypass capacitor C_f shunts feedback noise currents but retains a minimum gain of unity for the amplifier's noise voltage.

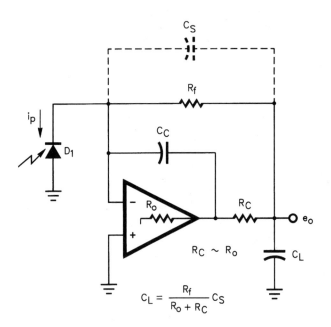

Figure 6.7 Addition of a low-pass filter in series with the amplifier output restricts noise bandwidth by using a decoupling phase compensation normally applied to contend with capacitance loading.

That unity-gain extension encompasses the higher-frequency range of the op amp, often producing a dramatic noise increase. Here, the low-pass filter continuously rolls off all noise signals of the circuit, uninterrupted by that unity-gain limit. The noise result resembles that of the previous composite amplifier with the low-pass filter here replacing the composite's integrator. Also, this filtering action remains within the feedback loop of the op amp, avoiding the added e_{id2} error described for the active filter alternative above.

Normally, adding a low-pass filter in the feedback path of an op amp compromises frequency stability. The filter's pole combines with that of the op amp, producing an unstable two-pole feedback. However in this case, the addition of capacitor C_C restores stability by decoupling the amplifier from the filter. At higher frequencies, C_C bypasses the overall feedback but does not diminish the noise reduction of the low-pass filter. For this bypass, C_C provides a lower-impedance feedback path at higher frequencies that replaces the primary path of R_C and R_f. Then, at higher frequencies, the op amp operates with a simple short-circuit feedback through C_C. Thus, the C_C bypass phase compensates the circuit as long as the amplifier exhibits unity-gain stability. Independent of this bypass, the low-pass filter components R_C and C_L continue to attenuate higher-frequency noise at the circuit's output. The circuit's phase compensation requirement combines with a noise-versus-bandwidth compromise to determine design equations for the circuit components.

6.4.2 Optimizing the decoupling noise-versus-bandwidth compromise

Up to a point, this phase-compensated photodiode amplifier reduces noise bandwidth without reducing signal bandwidth. Reducing the circuit's roll off frequency increases the noise reduction but also requires an increased phase compensation, making a new bandwidth limit significant. Then, two limits bound the signal bandwidth with the lower of the two prevailing. As with the composite alternative, a compromise equalizes the two limits to optimize the resulting noise-versus-bandwidth compromise. The first bandwidth limit results from the usual parasitic bypass of the converter's feedback, and the second results from the feedback bypass introduced by C_C. Frequency response plots guide the development of design equations that align the two limits while preserving frequency stability.

The combined stability and bandwidth requirements align the three pole frequencies produced by C_L, C_C, and C_S. Alignment of the first two ensures circuit stability and the third alignment optimizes bandwidth. For the associated analyses, Fig. 6.8 shows the photodiode amplifier's basic and modified A_{OL} response curves. There, the shaded area between the two represents the noise response removed by the phase compensation circuit. Load capacitance C_L produces the response modification by adding a second A_{OL} pole at $f_{pL} = 1/2\pi(R_o + R_C)C_L$, where R_o is the output resistance of the op amp. The modified A_{OL}' response truncates noise gain A_{ne}, shifting two characteristic frequencies of the A_{ne} noise response. As shown, intercept frequency f_i and crossover frequency f_c shift to their filtered counterparts at f_i' and f_c'. Continued increase of C_L moves the latter frequencies to ever lower levels, for greater noise reduction, but also with a potential bandwidth consequence.

Indirectly, the pole added at f_{pL} introduces the second bandwidth limit through a phase compensation requirement. This requirement leads to a compromise satisfied by a first pole-frequency alignment. Stable phase compensation requires that the C_C bypass take feedback control before the modified A_{OL}' response fully develops its two-pole

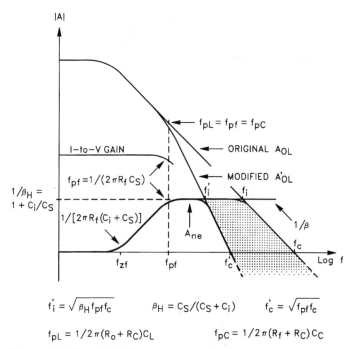

Figure 6.8 The added filter of Fig. 6.7 removes the shaded area of noise gain with a more aggressive, two-pole roll off but must initiate the roll off at a higher frequency to avoid a restriction of signal bandwidth.

roll off. Caution suggests increasing C_C to place the associated bypass frequency f_{pC} well before f_{pL}, but this would unnecessarily restrict signal bandwidth. Normally, the feedback pole at f_{pf} sets the signal or I-to-V bandwidth limit for higher-gain photodiode amplifiers. There, the stray capacitance bypass of the high feedback resistance shunts part of the input current away from that resistance. In Fig. 6.7, capacitance C_C also shunts the circuit's input current, producing a second bandwidth limit. In compromise, setting $f_{pC} = f_{pL}$ sufficiently counteracts the f_{pL} phase effects by retaining a 45° phase margin.

A third pole alignment defines the $f_{pC} = f_{pL}$ frequency described above. Aligning the new bandwidth limit at f_{pC} with the original limit at f_{pf} optimizes the noise-versus-bandwidth compromise. Making the circuit's two bandwidth limits coincident, $f_{pC} = f_{pf}$, only reduces signal bandwidth from f_{pf} to $BW_t = 0.64 f_{pf}$. Then, increasing C_L until $f_{pL} = f_{pC}$ maximizes the filtering noise reduction without compromising frequency stability. Thus, the $f_{pL} = f_{pC} = f_{pf}$ alignment retains stability, maximizes noise reduction, and avoids excessive bandwidth restriction.

The above three-frequency alignment defines design equations that aid the selection of circuit components. These equations provide the analytical starting point for the values of C_L and C_C given the practical constraints of R_o and R_C. First, the selection of R_C and C_L defines the filter's pole frequency, $f_{pL} = 1/2\pi (R_o + R_C)C_L$, with attention to output swing requirements. Resistor R_C must be large enough to decouple the amplifier from capacitance load C_L but not so large as to significantly limit output voltage swing under load current conditions. Typically, making $R_C \approx R_o$ meets these requirements, where R_o is the high-frequency value of the amplifier's output impedance. The complex frequency response of op amp output impedance would generally require empirical approximation of an R_o value, but simply assuming $R_o \approx 50\,\Omega$ typically suffices here. Later empirical tuning of C_C adjusts for the approximation error of this choice. Then, given the value for $R_C = R_o$, the value of C_L follows from equating $f_{pL} = 1/2\pi (R_o + R_C)C_L = f_{pf} = 1/2\pi R_f C_S$. Solving for C_L produces

$$C_L = \frac{R_f}{R_o + R_C} C_S$$

where C_S is the stray capacitance bypassing R_f. Typically, $C_S \approx 0.5$ pF. Next, the $f_{pC} = f_{pf}$ alignment, where $f_{pC} = 1/2\pi (R_f + R_C)C_C$ and $f_{pf} = 1/2\pi R_f C_S$, produces

$$C_C = \frac{R_f}{R_f + R_C} C_S \approx C_S$$

The above equation for C_C represents the starting point for this element's design value. Empirical tuning yields the final result, as required by the uncertainties of R_o and C_S. The amplifier output impedance, roughly modeled by R_o, and the variable stray capacitance C_S preclude definitive design equations. Accurate modeling of these quantities presents a greater task than the empirical bench testing and design tuning. For this tuning, driving the amplifier's noninverting input with a square wave and observing the output response reveals the circuit's step response characteristic. Then, adjusting the value of C_C removes any oscillation, moderates ringing, and limits overshoot without unnecessarily restricting signal bandwidth.

6.4.3 Noise analysis for the decoupling case

Evaluation of the resulting noise reduction requires just four adjustments to the original noise results developed in Sec. 5.3.1. Like the preceding composite photodiode amplifier, the filtering photodiode amplifier reduces the E_{noR}, E_{noi}, E_{noe4}, and E_{noe5} components of the output noise. Once again, E_{noR} and E_{noi} decrease due to the two-pole roll off now presented to the noise effects of the feedback resistor and the amplifier input noise current. Making $f_{pC} = f_{pf}$ places the circuit's two bandwidth limits at the same frequency, producing the two-pole roll off shown in the figure. This repeats the composite amplifier results described in Sec. 6.3.4, making

$$E_{noR} = \sqrt{KTR_f \pi f_{pf}} \quad \text{and} \quad E_{noi} = R_f \sqrt{\frac{q \pi f_{pf} I_{B-}}{2}}$$

The other two components affected, E_{noe4} and E_{noe5}, decrease both due to the steeper roll off and the reduced frequency range of the A_{ne} noise gain. These differences alter the E_{noe4} and E_{noe5} results developed for noise regions 4 and 5 of the original Fig. 5.5 analysis. Here, the $1/\beta$ intercept of Fig. 6.8 now terminates the A_{ne} plateau of the original region 4 at f_i' rather than f_i. Also, f_i' replaces f_i as the beginning of the previous region 5 noise roll off. Further, this roll off now initiates a two-pole roll rather than single-pole decline.

Geometric evaluation of the figure defines f_i', and rms evaluation of the two-pole effect completes the noise result adjustments for this decoupling case. As before, a modified listing summarizes the net results for a convenient reference to step-by-step noise analysis. First, geometric evaluation of the response curves defines f_i' in terms of known quantities for the case where $f_{pL} = f_{pf}$. Note the right triangles formed by the dashed f_{pf} indicator line with the A_{OL} and A_{OL}' responses. Comparison of these triangles defines a new crossover frequency f_c' that later reflects back to f_i'. For the original A_{OL} response, a single-pole roll off drops this response to the unity-gain axis at f_c, defining an endpoint for the associated triangle base. Similarly, the two-pole roll off of the A_{OL}' response defines f_c' as the corresponding endpoint. With twice the slope, the A_{OL}' response produces a triangle base with one-half the length. Then, given the logarithmic frequency scale of the plot, $\text{Log}(f_c') - \text{Log}(f_{pf}) = 0.5[\text{Log}(f_c) - \text{Log}(f_{pf})]$. Solving for f_c' yields $f_c' = \sqrt{f_{pf}f_c}$. Further examination of the A_{OL}' response translates this f_c' result into the desired f_i'. In its two-pole roll off region, this response approximates $A_{OL}' \approx f_c'^2/f^2$. At the f_i' intercept, the A_{OL}' and $1/\beta_H$ curves cross, making $A_{OL}' = f_c'^2/f_i'^2 = 1/\beta_H$, and solving for f_i' yields

$$f_i' = \sqrt{\beta_H f_{pf} f_c}$$

where $\beta_H = C_S/(C_i + C_S)$. Substituting this value of f_i' for f_i in the previous noise result for E_{noe4} adjusts that result for the reduced noise bandwidth imposed by f_i'.

The E_{noe5} result requires greater adjustment to account for the two-pole roll off of the modified A_{OL}' response. This accelerated roll off dramatically reduces the noise contribution of this highly significant region. Following the Sec. 5.2.3 discussion, the rms noise content of region 5 is described by

$$E_{noe5}^2 = \int_{f_i'}^{\infty} |A_{ne}e_{ni}|^2 \, df$$

This region represents the circuit's noise contribution from f_i' upward in Fig. 6.8. There, A_{ne} follows the A_{OL}' response, making $A_{ne} = A_{OL}' \approx f_c'^2/f^2$, where $f_c' = \sqrt{f_{pf}f_c}$ as described above. Then, $A_{ne} \approx f_{pf}f_c/f^2$. Also in this region, e_{ni} follows its floor level for $e_{ni} = e_{nif}$. Substituting these A_{ne} and e_{ni} results in the E_{noe5} integral produces

$$E_{noe5}^2 = \int_{f_i'}^{\infty} |f_{pf}f_c e_{nif}/f^2|^2 \, df = \frac{(f_{pf}f_c e_{nif})^2}{3 f_i'^3}$$

where $f_i' = \sqrt{\beta_H f_{pf} f_c}$ from above. Solving for E_{noe5} yields

$$E_{noe5} = \frac{(f_{pf}f_c)^{1/4}}{\sqrt{3}(\beta_H)^{3/4}} e_{nif}$$

where $f_{pf} = 1/2\pi R_f C_S$, and $\beta_H = C_S/(C_S + C_i)$.

Substituting the above E_{noR}, E_{noi}, E_{noe4}, and E_{noe5} results adjusts the noise analysis results of Sec. 5.3.1 for this filtering approach to noise reduction. Then, for Fig. 6.7:

$E_{noe1} = e_{nif}\sqrt{f_f \ln(f_f/f_1)}$, where e_{nif} is the e_{ni} noise floor level, and f_f is the $1/f$ noise corner frequency. As explained in Sec. 5.2.3, setting $f_1 \approx 0.01$ Hz avoids a theoretical contradiction and ensures analytical accuracy.

$E_{noe2} = e_{nif}\sqrt{f_{zf} - f_f}$, where $f_{zf} = 1/2\pi R_f(C_i + C_S)$, and $C_i = C_D + C_{id} + C_{icm}$. Here and in the E_{noe4} analysis, numerical evaluation can produce the square root of a negative number. Then, substitute zero for the result as explained in Sec. 5.3.1.

$E_{noe3} = (e_{nif}/f_{zf})\sqrt{\left(f_{pf}^3 - f_{zf}^3\right)/3}$, where $f_{pf} = 1/2\pi R_f C_S$, and C_S is the stray feedback capacitance.

$E_{noe4} = (1 + C_i/C_S)e_{nif}\sqrt{f_i' - f_{pf}}$, where $f_i' = \sqrt{\beta_H f_{pf} f_c}$, and $\beta_H = C_S/(C_i + C_S)$.

$E_{noe5} = (f_{pf}f_c)^{1/4}e_{nif}/\sqrt{3}(\beta_H)^{3/4}$, where f_c is the op amp's unity-gain crossover frequency.

$E_{noe} = \sqrt{E_{noe1}^2 + E_{noe2}^2 + E_{noe3}^2 + E_{noe4}^2 + E_{noe5}^2}$, the total rms output noise resulting from the spectral noise voltage e_{ni} of the amplifier input.

$E_{noR} = \sqrt{KTR_f \pi f_{pf}}$, where $f_{pf} = 1/2\pi R_f C_S$. For applications that add a feedback capacitance C_f, replace C_S with $C_f + C_S$.

$E_{noi} = R_f\sqrt{q \pi f_{pf} I_{B-}/2}$.

$E_{no} = \sqrt{E_{noR}^2 + E_{noi}^2 + E_{noe}^2}$.

References

1. J. Graeme, "Filtering Cuts Noise in Photodiode Amplifiers," *Electronic Design*, November 7, 1994, p. 9.
2. R. Burt and R. Stitt, "Circuit Lowers Photodiode-Amplifier Noise," *EDN*, September 1, 1988, p. 203.

Chapter

7

High-Gain Photodiode Amplifiers

For the basic photodiode amplifier, just making the feedback resistance large produces high gain. However, the very high resistances often required introduce other performance limitations. For those cases, alternative methods described here provide reduced offset or increased bandwidth but produce some compromise to noise performance. First, reduced dc offset results from replacing the circuit's feedback resistor with a tee network, as previously described in Sec. 2.2.2. Here, noise analysis of this alternative reveals that judicious design largely avoids the noise increase commonly encountered with the feedback tee.

Increased bandwidth results from supplemental gain provided by three other circuit alternatives. For a given transresistance, the supplemental gain reduces the feedback resistance required, and improves the bandwidth, by a factor equal to that gain. The most straightforward implementation simply adds a voltage amplifier following the conventional current-to-voltage converter. A second reduces the circuit complexity by making one op amp serve both the current-to-voltage and voltage-gain functions. In this case, a bootstrap connection for the photodiode removes the signal swing from the photodiode, permitting this dual function without sacrificing other performance. Current output along with increased bandwidth result from replacing the added voltage gain with current gain. There, a bootstrap connection avoids the oscillation encountered with straightforward application of an op amp current amplifier. For each of the tee and supplemental gain alternatives, optimization of a noise compromise guides component selection for the circuit.

Performance optimization with each also requires attention to the amplifier's physical construction. There, selecting a low-capacitance conversion resistor and using low-capacitance assembly methods minimize the bandwidth-limiting parasitic capacitance. With printed circuit board assemblies, removing the board's ground plane under the conversion resistor and mounting this resistor on standoffs reduce the assembly-related parasitic. Note that the standoff at the circuit input should be insulated to reduce leakage currents; otherwise dc offset could increase. Further, the standoff mountings should be rigid to avoid the potential for parasitic noise induced in the resistor by microphonic effects.

7.1 Using a Feedback Tee Network

A tee network significantly reduces the dc offset error of high-gain current-to-voltage converters, as described in Sec. 2.2.2. There, replacing the high-value feedback resistor with a tee, composed of lower-resistance elements, permits the offset reduction of a compensation resistance without simultaneously increasing offset through increased photodiode leakage current. However, the tee also amplifies the op amp's input offset voltage, producing the Sec. 2.2.2 design limit

$$\frac{R_1}{R_2} \leq \frac{I_{B-} - I_{B+} + I_L}{10 V_{OS}} R_{\text{feq}} - 1$$

Circuit noise imposes two other limits for this R_1/R_2 selection. Feedback tee networks have a reputation for high noise gain, suggesting tee avoidance for the characteristically noise-stressed photodiode amplifier. However, closer examination of the tee alternative reveals that optimizing its design avoids noise degradation for those photodiode amplifiers serving large-area diodes.[1] There, the diode capacitance makes noise gain peaking and the amplifier noise voltage the dominant noise effects. The tee network increases the effect of this voltage noise and the effect of resistance noise. However, design equations guide the tee selection to keep these increases in the background of the initial gain peaking dominance. These equations first limit the increase in the voltage noise to the less-significant, lower-frequency range. Next, the equations prevent resistance noise from becoming the dominant effect. Then, the tee network produces the desired reduction in dc offset for only a small increase in net noise.

7.1.1 The gain and noise produced by a feedback tee

Figure 7.1 models the tee-feedback photodiode amplifier for analysis. There, a current source i_p and a capacitance C_i account for the effects of the photodiode. Capacitance C_i also includes the differential and

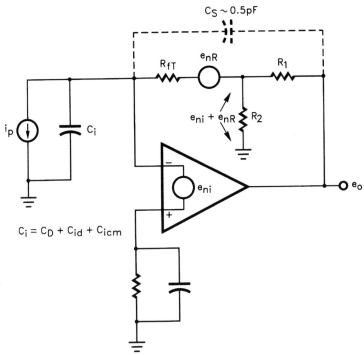

Figure 7.1 Replacing the feedback resistor with a tee network increases the low-frequency gain for noise sources e_{ni} and e_{nR}, but judicious component selection prevents a gain increase in the dominant high-frequency range.

common-mode input capacitances of the op amp, making $C_i = C_D + C_{id} + C_{icm}$. Noise sources e_{ni} and e_{nR} represent the noise voltages of the amplifier and the feedback resistor. For current-to-voltage conversion, the tee network acts as an equivalent feedback resistance of

$$R_{\text{feq}} = R_1 + \left(1 + \frac{R_1}{R_2}\right) R_{fT} \approx \left(1 + \frac{R_1}{R_2}\right) R_{ft} \quad \text{for the high-gain case}$$

Here, the feedback attenuation of the R_1, R_2 divider multiplies the effective resistance of R_{fT} by the inverse of the divider ratio. This inverse ratio, $1 + R_1/R_2$, characterizes the feedback action of the tee and reappears numerous times in the performance analyses that follow. In response to diode current i_p, this equivalent resistance makes the signal output $e_o = i_p R_{\text{feq}} = i_p (1 + R_1/R_2) R_{fT}$.

Noise adds to this ideal output signal and, true to its reputation, the feedback tee amplifies the e_{ni} and e_{nR} noise voltages of the figure. Comparison of noise results for the basic photodiode amplifier and the tee alternative defines design equations that make this noise amplification negligible. These equations prevent the added amplification from overriding the initial dominance of noise gain peaking. This peaking amplifies e_{ni} at higher frequencies and, where peaking dominates noise performance, the tee alternative can reduce offset without materially degrading noise performance. The noise gain A_{ne} characterizes this peaking, and Sec. 5.1.2 defines this gain for the basic photodiode amplifier as

$$A_{ne} = \frac{1 + R_f(C_i + C_S)s}{1 + R_f C_S s}$$

This gain begins at unity at low frequencies and then experiences a response zero at $f_{zf} = 1/2\pi R_f(C_i + C_S)$ followed by a pole at $f_{pf} = 1/2\pi R_f C_S$. Together, f_{zf} and f_{pf} produce an A_{ne} plateau level of $1 + C_i/C_S$.

Analysis of Fig. 7.1 defines the corresponding noise gain for the tee alternative. At low frequencies, where the circuit capacitances have no influence, the tee amplifies the noise signals e_{ni} and e_{nR} with the same gain. Neglecting i_p for noise analysis, a dc loop evaluation shows that feedback produces the combined noise voltage $e_{ni} + e_{nR}$ across R_2. To do so, the feedback action develops an output noise voltage $e_{no} = (1 + R_1/R_2)(e_{ni} + e_{nR})$. This reflects a low-frequency noise gain for the tee of $1 + R_1/R_2$ as compared with the unity gain of the basic current-to-voltage converter. For e_{nR} this noise gain simply continues until rolled off by the feedback bypass capacitance. However, e_{ni} again reacts with C_i to produce noise gain peaking, and the added $1 + R_1/R_2$ gain of the tee potentially increases this effect. Separate analyses for the two noise sources produce design guidelines that avoid a significant noise increase due to the added noise gain.

First consider the overall noise effect of e_{ni}. Figure 7.2 displays the A_{ne} noise gain response and the tee's range of A_{neT} alternatives that limit the noise increase. There, the dashed A_{ne} curve portrays the noise gain response of the basic photodiode amplifier with its characteristic zero at f_{zf} and pole f_{pf}. The solid A_{neT} curve displays the upper limit that potentially permits offset reduction without significantly increasing noise. At higher frequencies, this solid A_{neT} curve overlaps and follows the original A_{ne} in the peaking region that dominates the e_{ni} noise effect. Under this condition, the tee does not increase high-frequency noise gain. At lower frequencies, a dotted response portrays an intermediate A_{neT} result to illustrate the effect of the tee upon the response zero f_{zf}. For consistency, all three of the noise gain responses represent the same current-to-voltage gain with R_{feq} of the tee examples equal to R_f of the basic case.

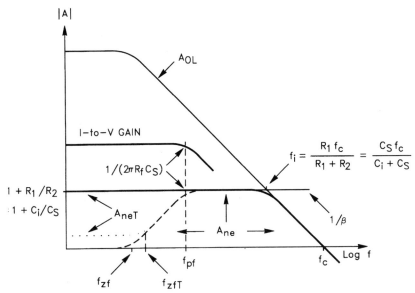

Figure 7.2 The Fig. 7.1 tee increases noise gain from A_{ne} to A_{neT}, but careful design limits the increase to the less-significant low-frequency range of the circuit response.

In the figure, the feedback tee increases the low-frequency level of A_{ne} but also moves its response zero to a higher frequency, avoiding any increase in the dominant noise gain peak. With careful design, the inherent noise increase of the tee remains in the background of this dominant effect. The higher-frequency zero of the tee case, f_{zT}, results from the lower net resistance presented by the tee to the amplifier's input circuit. The dashed curve of the basic circuit begins its frequency response at unity gain due to the single resistor feedback of the basic photodiode amplifier. This simple feedback presents the high resistance of R_f to the amplifier's input circuit, producing the response zero $f_{zf} = 1/2\pi R_f(C_i + C_S)$. With the tee, the noise gain A_{neT} begins its frequency response at a level of $1 + R_1/R_2$, rather than the unity level of the basic case. This increases the voltage noise effect, but careful design restricts this increase to the less-significant, lower-frequency range of the A_{ne} response.

Counteracting this low-frequency increase, the tee of Fig. 7.1 presents a far lower resistance to the input circuit, resulting in a higher-frequency A_{neT} zero at f_{zT}. For a given current-to-voltage gain, the tee produces an equivalent feedback resistance, $R_{feq} = R_f$, but this R_{feq} resistance does not react with the circuit's input capacitance to produce f_{zfT}. Seen from the amplifier's inverting input, the feedback network presents a resistance of $R_{fT} + R_1 \| R_2 \approx R_{fT}$

to this capacitance. Here, the return of R_1 to the low impedance of the amplifier output effectively connects it to ground, placing it in parallel with R_2. This input-referred feedback resistance breaks with the circuit's capacitances at $f_{zfT} = 1/2\pi R_{fT}(C_i + C_S)$. Comparing this result with the above f_{zf} expression shows that the tee moves the onset of gain peaking to a higher frequency, as determined by R_{fT} rather than R_f. Following f_{zfT}, capacitance bypass of the feedback path again reacts with the net resistance of the current-to-voltage gain, producing a pole at $f_{pfT} = 1/2\pi R_{feq}C_S = f_{pf}$. Summarizing these effects, the noise gain for the tee alternative becomes

$$A_{neT} = \left(1 + \frac{R_1}{R_2}\right) \frac{1 + R_{fT}(C_i + C_S)s}{1 + R_{feq}C_S s}$$

Here, different resistances control the zero and pole of A_{neT}, adding an extra degree of design freedom. For the basic current-to-voltage converter before, the same resistance R_f controls both the zero and pole of the corresponding noise gain.

7.1.2 Optimizing the tee's noise voltage response

Examination of the Fig. 7.2 response plots reveals the opportunity presented by the tee's added degree of freedom. For a given R_{feq}, increasing the $1 + R_1/R_2$ multiplier of the tee reduces the required value of R_{fT} and the resulting dc offset error as described in Sec. 2.2.2. Increasing this multiplier raises A_{neT} in the lower-frequency region of the response but also moves f_{zf} to a higher-frequency f_{zfT} and leaves the pole at f_{pf} unchanged. Thus, increasing the $1 + R_1/R_2$ multiplier simply walks the dotted A_{neT} curve up the dashed A_{ne} curve. This change only increases noise gain in the less-significant, low-frequency range as long as this dotted response remains within the bounds of the solid A_{neT} response. Then, the tee produces no A_{ne} change above the frequency f_{pf} that marks the beginning of the peak A_{ne} level.

The higher-frequency range of the gain peak spans the dominant spectrum of the circuit's frequency response and largely determines the circuit's response to the e_{ni} noise source. At first, visual evaluation of relative areas under the noise gain curves fails to convey this noise dominance. At lower frequencies, the large area added by the A_{ne} to A_{neT} conversion would seem to greatly affect the noise result. However, the logarithmic nature of the curve's frequency axis greatly exaggerates the noise significance of the curves at low frequencies. In practice, this area increase adds little to the net noise result.

However, excessive increases in the tee's $1+R_1/R_2$ gain potentially increases the higher-frequency noise gain as well. The solid A_{neT} curve shown defines an upper limit bounding the tee design. This curve raises the mean level of the noise gain to the peak level of the original A_{ne} curve, but no more. Any further increase would increase noise gain in the dominant high-frequency region, significantly degrading noise performance. To define this boundary, the selection of the $1+R_1/R_2$ multiplier produces pole-zero cancellation for the noise gain response. Then, the tee moves the f_{zfT} zero up the A_{ne} response until it cancels the pole at f_{pf}, as represented by the solid A_{neT} curve. Further examination of the previous A_{neT} equation

$$A_{neT} = \left(1 + \frac{R_1}{R_2}\right) \frac{1 + R_{fT}(C_i + C_S)s}{1 + R_{feq}C_S s}$$

shows that pole-zero cancellation produces a constant noise gain of $A_{neT} = 1 + R_1/R_2$.

This cancellation requires that $R_{fT}(C_i + C_S) = R_{feq}C_S$, where $R_{feq} = (1+R_1/R_2)R_{fT}$. Mathematical manipulation of this requirement reduces it to the simple expression $R_1/R_2 = C_i/C_S$, defining an upper limit for the R_1/R_2 ratio. Thus, limiting the noise increase due to the tee's amplification of e_{ni} requires that

$$\frac{R_1}{R_2} \leq \frac{C_i}{C_S}$$

where $C_i = C_D + C_{id} + C_{icm}$. This is the second limit equation for the tee-feedback current-to-voltage converter used with large area photodiodes. Verbally expressed, this limit equation states that the resistor ratio of the feedback tee's voltage divider should not exceed the ratio of the circuit's input and feedback capacitances. For the above limit equation, the amplifier and photodiode define C_i, but stray capacitance C_S remains a function of the actual application conditions. Typically, $C_S \approx 0.5$ pf.

7.1.3 Optimizing the tee's resistance noise response

The tee network also amplifies the resistor noise e_{nR}. This noise amplification potentially overrides the controlled noise increase described above, but imposing a third limit to the tee's amplification prevents the override. For a given application requirement, a trial design of the basic photodiode amplifier solution first determines whether or not e_{ni} initially dominates noise performance. There, comparison of the net rms noise effects reveals the dominant source for a given application. Where e_{ni} and noise gain peaking dominate, the tee offers offset reduction without a significant noise compromise.

Noise analysis for the tee alternative follows from that of the basic current-to-voltage converter presented in Chap. 5. There, analyses convert the effects of the e_{ni}, i_{ni}, and e_{nR} noise effects into the rms output noise components E_{noe}, E_{noi}, and E_{noR}. These fundamental noise components result from the amplifier's input voltage and current noise sources and from the resistor's noise. Together, these components combine to form the total output noise

$$E_{no} = \sqrt{E_{noe}^2 + E_{noi}^2 + E_{noR}^2}$$

Selecting an FET input amplifier normally makes the E_{noi} component negligible in the above summation. Then, as long as $E_{noe} \geq 3E_{noR}$, E_{noe} effectively dominates the E_{no} noise summation. This rms summation squares its component results before adding them, and there $E_{noe} \geq 3E_{noR}$ produces $E_{noe}^2 \geq 9E_{noR}^2$. When this condition exists for the basic circuit, resistor noise adds little to the final result, allowing for an E_{noR} increase in the conversion to a tee feedback.

Analysis defines the resistance noise increase of the tee by extrapolating the E_{noR} result of the basic case. As described with Fig. 7.1, the tee amplifies the e_{nR} noise voltage by a gain of $1 + R_1/R_2$ instead of the unity gain of the basic case. Modifying the basic E_{noR} result of Sec. 5.3.1 for the tee produces $E_{noRT} = (1 + R_1/R_2)\sqrt{2KTR_{fT}\pi \text{BW}_t}$, where $\text{BW}_t = 1/2\pi R_{feq}C_S$. This result assumes the typical case where $R_{fT} \gg R_1$ or R_2, making the noise of R_{fT} the dominant noise source for the resistors of the tee. The tee's added gain makes E_{noRT} larger than the original E_{noR}, but the accompanying replacement of R_f with a smaller R_{fT} limits the increase. For a given $R_{feq} = (1 + R_1/R_2)R_{fT}$, solving for $R_{fT} = R_{feq}/(1 + R_1/R_2)$ shows that increasing the tee's $(1 + R_1/R_2)$ gain permits an equal reduction in R_{fT}. However, the E_{noRT} noise varies in direct proportion to the $(1 + R_1/R_2)$ gain but only in square root proportion to R_{fT}. Substituting $R_{fT} = R_{feq}/(1 + R_1/R_2)$ in the E_{noRT} expression produces $E_{noRT} = \sqrt{1 + R_1/R_2}\sqrt{2KTR_{feq}\pi \text{BW}_t}$. Thus, for a given $R_{feq} = R_f$, replacing R_f with the tee increases E_{noR} by the factor $\sqrt{1 + R_1/R_2}$ to $E_{noRT} = \sqrt{1 + R_1/R_2}E_{noR}$.

This $\sqrt{1 + R_1/R_2}$ factor and the $E_{noe} \geq 3E_{noR}$ guideline produce the third limit equation for the tee's gain. Depending upon the E_{noe}/E_{noR} ratio of the earlier trial design, the circuit may or may not permit conversion to a tee without a significant noise increase. Where $E_{\text{noe}} > 3E_{\text{noR}}$ for the basic case, the increased E_{noRT} can be accepted up to a point as guided by the specific $E_{\text{noe}}/E_{\text{noR}}$ ratio of the trial design. A ratio of $E_{\text{noe}}/E_{\text{noR}} \leq 3$ indicates a lack of or marginal E_{noe} dominance, and replacing R_f with the tee would significantly increase noise. However, $E_{\text{noe}}/E_{\text{noR}} > 3$ indicates a surplus noise dominance, permitting the tee replacement of R_f. Then, adding the tee and increasing E_{noR} until $E_{\text{noe}}/E_{\text{noRT}} = 3$ optimizes the offset-reduction-versus-noise compromise. Substituting the $E_{\text{noRT}} = \sqrt{1 + R_1/R_2} E_{\text{noR}}$ expression into this condition and solving for R_1/R_2 produces

$$\frac{R_1}{R_2} \leq \left(\frac{E_{\text{noe}}}{3E_{\text{noR}}}\right)^2 - 1$$

where E_{noe} and E_{noR} represent the numerical noise values resulting from the trial design of the basic case. This is the third limit equation for the tee-feedback current-to-voltage converter used with large-area photodiodes.

Combined evaluation of the tee's three R_1/R_2 gain limit equations defines the tee design for a given case. The three equations prevent an offset increase due to V_{OS}, limit the increase in E_{noe}, and prevent noise dominance by E_{noR}. Evaluation of the limit equations produces three steps for the tee design. First, the limit equation just above must produce a positive value for R_1/R_2, indicating a surplus E_{noe} dominance and an opportunity for offset reduction without significantly increasing the effect of E_{noR}. Next, this positive value must satisfy the previous condition $R_1/R_2 \leq C_i/C_S$ to avoid significantly increasing E_{noe}. Finally, the R_1/R_2 ratio must meet the condition $R_1/R_2 \leq (I_L + I_{B-} - I_{B+})R_{\text{feq}}/10V_{\text{OS}} - 1$ to limit the V_{OS} effect upon offset. Section 2.2.2 develops this offset limit equation. For a given application, the smallest R_1/R_2 result from the three equations defines the offset-versus-noise optimum.

After determining R_1/R_2, choose R_1 for a convenient resistance value that does not excessively load the amplifier output. This choice simultaneously defines R_2 through the R_1/R_2 ratio selected. Given R_1 and R_2, the value of R_{fT} follows from the application's required gain as set by $R_{\text{feq}} \approx (1 + R_1/R_2)R_{\text{fT}}$. Solving for R_{fT}

$$R_{\text{fT}} = \frac{R_{\text{feq}}}{1 + R_1/R_2}$$

This concludes the tee design.

7.1.4 Noise analysis for the tee case

Replacing R_f with the tee produces numerous changes to the noise analysis results for the photodiode amplifier. Changes occur in all three of the fundamental rms output noise results E_{noe}, E_{noR}, and E_{noi}. These represent the effects of the amplifier input noise voltage e_{ni}, the primary feedback resistor R_{fT}, and the amplifier input current noise i_{ni}, respectively. The last subscript character of the rms labels identifies the effects with the corresponding sources. In spite of the great change, simple modifications adapt the previous noise results to this tee case, avoiding the repeat of a lengthy analysis. These modifications only require inclusion of the tee's low-frequency noise gain, $1 + R_1/R_2$, and a change in one frequency limit for the E_{noe} analysis.

The simplest changes occur in the E_{noR} and E_{noi} results, where adding the new low-frequency noise gain and replacing R_f with R_{fT} adjusts the earlier results to the tee feedback case. As described above, the tee increases the low-frequency noise gain from unity to $1 + R_1/R_2$. However, the tee also substitutes the much smaller R_{fT} for R_f for the circuit's dominant resistance, reducing noise. For the resistance noise, the preceding section shows that these changes make $E_{noR} = (1 + R_1/R_2)\sqrt{2KTR_{fT}\pi BW_t}$ where $BW_t = 1/2\pi R_{feq}C_S$. Similarly, the current noise component E_{noi} results from a noise voltage developed on R_{fT} that receives the same gain. Adding this gain and replacing R_f with R_{fT} converts the Sec. 5.3.1 noise result to $E_{noi} = (1 + R_1/R_2)R_{fT}\sqrt{2q\pi BW_t I_{B-}}$. Noting that $R_{fT} = R_f/(1+R_1/R_2)$ and substituting this expression reveals the actual changes in noise performance for E_{noR} and E_{noi}. Then, $E_{noR} = \sqrt{(1 + R_1/R_2)(2KTR_f\pi BW_t)}$ and $E_{noi} = R_f\sqrt{q\pi BW_t I_{B-}}$. From these expressions, the added noise gain of the tee increases E_{noR} by $\sqrt{1 + R_1/R_2}$ and leaves E_{noi} unchanged.

The tee's low-frequency noise gain also alters E_{noe} through the modified A_{ne} amplification of the amplifier's input noise voltage e_{ni}. Previously, Fig. 5.5 illustrated the basic e_{noe} noise density curve that produces E_{noe} from the $A_{ne}e_{ni}$ combination. Conversion to the tee feedback leaves the general shape of this curve unchanged, as shown in Fig. 7.3. There, the e_{noe} curve repeats the five regions that segment the rms noise analysis for simplified development of the total E_{noe} noise result. Region 1 still displays the $1/f$ noise influence of e_{ni} followed by its noise floor influence in region 2. Region 3 displays the rise of the noise gain peaking beginning with a response zero and ending with a pole. Then, the region 4 plateau and region 5 roll off duplicate the previous regional responses. However, this curve does differ from the basic one in regions 1 through 3. In regions 1 and 2, the increased low-frequency noise gain raises the curve by a factor of

High-Gain Photodiode Amplifiers 139

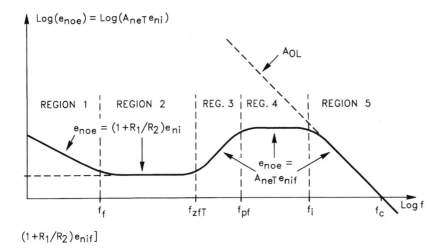

$$E_{noe} = \sqrt{E_{noe1}^2 + E_{noe2}^2 + E_{noe3}^2 + E_{noe4}^2 + E_{noe5}^2}$$

$f_{zfT} = 1/2\pi R_{fT}(C_i + C_S)$ $f_{pf} = 1/2\pi R_{feq}C_S$ $f_i = f_c C_S/(C_i + C_S)$

Figure 7.3 The tee network raises the base levels of the noise response for regions 1 and 2 of the original Fig. 5.5 noise response.

$1 + R_1/R_2$. In region 3, a new starting point at the level $(1+R_1/R_2)e_{nif}$ rather than e_{nif} begins the noise gain rise at f_{zfT} rather than f_{zf}. Otherwise, the curve remains the same as for the basic case, and simple modifications easily translate the basic results to this tee case.

The listing below incorporates the preceding changes to the original listing of Sec. 5.3.1 for the tee feedback case. Noise analysis then follows the steps:

$E_{noe1} = (1 + R_1/R_2)e_{nif}\sqrt{f_f \ln(f_f/f_1)}$, where e_{nif} is the e_{ni} noise floor level, and f_f is the $1/f$ noise corner frequency. As explained in Sec. 5.2.3, setting $f_1 \approx 0.01$ Hz avoids a theoretical contradiction while retaining analytical accuracy.

$E_{noe2} = (1 + R_1/R_2)e_{nif}\sqrt{f_{zfT} - f_f}$, where $f_{zfT} = 1/2\pi R_{fT}(C_i + C_S)$, and $C_i = C_D + C_{id} + C_{icm}$. Here and in the E_{noe4} analysis, numerical evaluation can produce the square root of a negative number. Then, substitute zero for the result as explained in Sec. 5.3.1.

$E_{noe3} = (1 + R_1/R_2)(e_{nif}/f_{zfT})\sqrt{\left(f_{pf}^3 - f_{zfT}^3\right)/3}$, where $f_{pf} = 1/2\pi R_{feq}C_S$, $R_{feq} = (1 + R_1/R_2)R_{fT}$, and C_S is the stray feedback capacitance.

$E_{noe4} = (1 + C_i/C_S)e_{nif}\sqrt{f_i - f_{pf}}$, where $f_i = f_c C_S/(C_i + C_S)$, and f_c is the unity-gain crossover frequency of the op amp.

$E_{noe5} = (e_{nif}f_c)\sqrt{1/f_i}$.

$E_{noe} = \sqrt{E_{noe1}^2 + E_{noe2}^2 + E_{noe5}^2}$, the total rms output noise resulting from the spectral noise voltage e_{ni} of the amplifier input.

$E_{noR} = (1 + R_1/R_2)\sqrt{2KTR_{fT}\pi BW_t}$, where $BW_t = 1/2\pi R_{feq}C_S$. For applications that add a feedback capacitance C_f, replace C_S with $C_f + C_S$.

$E_{noi} = (1 + R_1/R_2)R_{fT}\sqrt{q\pi BW_t I_{B-}}$

$E_{no} = \sqrt{E_{noR}^2 + E_{noi}^2 + E_{noe}^2}$

As described before, special cases potentially deliver the square root of a negative number in the calculation of the E_{noe2} and E_{noe4} components. These components cover flat regions that certain circuit conditions eliminate in Fig. 7.3. The condition $f_{zfT} \le f_f$ eliminates region 2, and $f_i \le f_{pf}$ eliminates region 4. Very large $R_f C_D$ time constants can produce one or both of these conditions. Then, the eliminated region or regions produce zero noise contribution. However, to cover all possible cases the segmented E_{noe} analysis must include the potential region 2 and region 4 noise contributions. Numerical evaluations of the E_{noe2} and E_{noe4} equations easily identify these special cases by delivering the square root of a negative number. As before, simply substitute zero for such analysis results.

7.2 Adding a Voltage Amplifier

Adding voltage gain to the basic current-to-voltage converter increases bandwidth for the high-gain case. The added gain permits a corresponding reduction in the converter's feedback resistance without reducing the circuit's net transimpedance. Then, the decreased resistance increases the bandwidth limit imposed by parasitic capacitance bypass. However, the combination of current-to-voltage conversion and voltage gain degrades noise performance, producing a bandwidth-versus-noise compromise. Judicious design choices make the bandwidth increase exceed the noise degradation, and analysis yields design equations that optimize this compromise. Two alternatives add this voltage gain. The first simply adds a voltage amplifier after the current-to-voltage converter, and the second adds voltage gain to the converter itself. The first alternative, described here, offers a greater

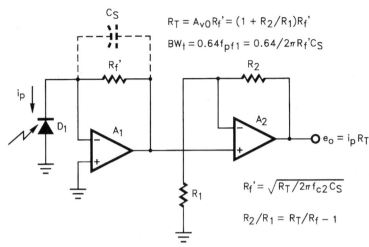

Figure 7.4 Adding a voltage amplifier after the current-to-voltage converter produces a net transresistance R_T with a smaller R_f that is less susceptible to the bandwidth shunting of C_S.

bandwidth improvement, and the second offers a simpler circuit. A third alternative, explained later, adds current rather than voltage gain for similar bandwidth improvement and a current rather than voltage output.

In Fig. 7.4, a second amplifier supplies voltage gain following the current-to-voltage converter, developing a net input-to-output transresistance R_T. Here, $R_T = A_{v0}R_f'$, where $A_{v0} = 1 + R_2/R_1$ is the lower-frequency value of the gain A_v supplied by A_2. Then, for a given value of R_T, this option reduces R_f' by a factor equal to A_{v0}. For high values of R_T, this resistance reduction decreases the resistor's sensitivity to parasitic shunting and improves bandwidth by the same A_{v0} factor. However, a second limit to signal bandwidth bounds the attainable improvement due to the finite bandwidth of the added amplifier. The resulting compromise suggests an optimization that increases bandwidth up to this second limit. This simultaneously optimizes the circuit's bandwidth-versus-noise performance as described below.

7.2.1 Optimizing the two-amplifier bandwidth-versus-noise compromise

While an obvious alternative, the two-op-amp solution's overall effects upon bandwidth and noise are not quite so obvious. The signal bandwidth limitations of both amplifiers now bound the upper end of the attainable circuit bandwidth. In addition, the input noise of the second amplifier adds to the overall circuit noise with a net effect dependent upon both the gain added and the resultant bandwidth. Figure 7.5 illustrates the bandwidth and noise sensitivities to A_{v0} with the BW_t and E_{no} curves. These curves represent the net transimpedance bandwidth and the net output noise voltage resulting for the two-amplifier case. Fortuitously, the bandwidth limit of A_2 makes the BW_t and E_{no} curves peak at the same optimum value of A_{v0o}. Above A_{v0o}, this bandwidth limit rolls off both signal and noise, equalizing their bandwidths, as sought in Chap. 6. As described there, the basic current-to-voltage converter often provides greater bandwidth for the amplifier voltage noise than for the photodiode signal. The two-amplifier circuit here removes that difference, using A_2's bandwidth limit as a noise filter.

The coincident peaks at A_{v0o} present a performance optimum due to the different slopes of the BW_t and E_{no} curves. Examination of these slopes reveals the bandwidth-versus-noise optimum. Prior to A_{v0o}, both BW_t and E_{no} rise with increasing A_{v0}, suggesting a stalemate in the bandwidth-versus-noise compromise. However, closer examination of the BW_t and E_{no} origins shows differing rates of increase. Below A_{v0o}, the BW_t curve rises with a greater slope than E_{no}, increasing the ratio BW_t/E_{no} up to the A_{v0o} optimum. The slope difference results from the differing BW_t and E_{no} sensitivities to A_{v0}. Increasing A_{v0} permits a proportionate decrease in R_f', directly increasing $BW_t = 1/2\pi R_f' C_S = A_{v0}/2\pi R_T C_S$. This direct proportionality, $BW_t \propto$

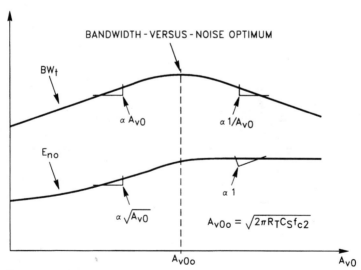

Figure 7.5 Increasing the voltage gain A_{v0} in Fig. 7.4 increases bandwidth at a greater rate than it increases noise up to an optimum value of A_{v0o}.

A_{v0}, defines the BW_t-versus-A_{v0} slope up to the A_{v0o} optimum. Beyond A_{v0o}, continued decrease in R_f' further reduces the $1/2\pi R_f'C_S$ limit, but there the finite gain-bandwidth product of A_2 takes control of the circuit's bandwidth. As analyzed later, this limit makes $BW_t = f_{c2}/A_{v0}$, where f_{c2} is the unity-gain crossover frequency of A_2. This rolls off the BW_t curve, reversing its slope from $BW_t \propto A_{v0}$ to $BW_t \propto 1/A_{v0}$ at the A_{v0o} optimum.

Similar analyses of the E_{no} curve defines its varying slope. Below A_{v0o}, increasing A_{v0} also increases E_{no}, but more gradually than BW_t. In this region, first noise gain peaking and then resistor noise control the slope of the E_{no} curve. The peaking characteristically controls the slope at the low end of the A_{v0} axis. There, the amplification of amplifier input noise voltage dominates E_{no}. Increasing A_{v0} reduces R_f', and the resulting noise gain peaking, but further amplifies the noise signal appearing at the A_1 output. The net result of these counteracting effects produces a slight, initial upward slope in the E_{no}-versus-A_{v0} curve. At somewhat higher A_{v0} levels, the corresponding reduction in R_f' removes the gain peaking dominance but replaces it with the combined effect of R_f' and A_{v0}. There, the increase in A_{v0} decreases R_f' in direct proportion through the relationship $R_f' = R_T/A_{v0}$. This decreases the resistor's noise voltage, as described by $e_{nR} = \sqrt{4KTR_f'}$, and would make $E_{no} \propto \sqrt{R_f'}$ for $E_{no} \propto 1/\sqrt{A_{v0}}$. However, the A_{v0} increase also amplifies the e_{nR} noise voltage in direct proportion, making $E_{no} \propto A_{v0}/\sqrt{A_{v0}} = \sqrt{A_{v0}}$. This final proportionality becomes the slope of the E_{no} curve preceding the A_{v0o} optimum.

Beyond this point, the E_{no} curve flattens due to the counteracting effects of the increasing gain versus the declining R_f' and bandwidth. The increasing gain tends to make $E_{no} \propto A_{v0}$, but the declining resistance and bandwidth cancel this effect. The rms noise of the resistor, $E_{nR} = \sqrt{4KTR_f'BW_t}$, exhibits square root dependencies upon both R_f' and BW_t. Above A_{v0o}, both $R_f' = R_T/A_{v0}$ and $BW_t = f_{c2}/A_{v0}$ display inverse proportionalities to A_{v0}. The combined effects make $E_{no} \propto A_{v0}/\sqrt{(A_{v0})^2} = 1$. Thus, above A_{v0o}, E_{no} becomes a constant with zero slope.

Comparing the slopes of the BW_t and E_{no} curves confirms the bandwidth-versus-noise optimum at A_{v0o}. Below this gain, the $BW_t \propto A_{v0}$ and the $E_{no} \propto \sqrt{A_{v0}}$ proportionalities make $BW_t/E_{no} \propto \sqrt{A_{v0}}$, and increasing A_{v0} improves the compromise. Above A_{v0o}, the $BW_t \propto 1/A_{v0}$ and the $E_{no} \propto 1$ proportionalities make $BW_t/E_{no} \propto 1/A_{v0}$, and increasing A_{v0} degrades the compromise. Thus, the value A_{v0o} that produces the coincident peaks of the BW_t and E_{no} curves also optimizes the circuit's bandwidth-versus-noise compromise.

7.2.2 Designing for the bandwidth-versus-noise optimum

To quantify this A_{v0o} value, consider the bandwidth curve BW_t and the conditions that optimize this characteristic. As A_{v0} increases, bandwidth initially increases in direct proportion due to the corresponding decrease in R_f'. There, the decreasing R_f' increases the frequency of the A_1 response roll off produced by stray capacitance. However, the added demands of the voltage gain on A_2 eventually make this amplifier's bandwidth the controlling factor, rolling off the BW_t-versus-A_{v0} curve. For a given value of R_T, an optimum value A_{v0o} produces a peak bandwidth like that shown. That peak occurs when the parasitic-limited bandwidth of A_1 and the gain-limited bandwidth of A_2 coincide. Parasitic or stray feedback capacitance C_S sets the A_1 bandwidth limit at $f_{pf1} = 1/2\pi R_f' C_S$. Further analysis defines the A_2 limit.

Feedback conditions similar to those described in Chap. 3 set the A_2 bandwidth at f_{i2}, the intercept frequency of this amplifier's $1/\beta$ curve with its A_{OL} response. Figure 7.6 illustrates this intercept at the crossing point of the $1/\beta_2$ and A_{OL2} curves. There, $1/\beta_2$ represents the feedback demand for gain, and A_{OL2} represents the amplifier gain available to supply this demand. At frequencies below the f_{i2} intercept, the available gain exceeds the feedback demand, and the excess or loop gain makes the closed-loop A_v curve follow the ideal level of its dc value, A_{v0}. At frequencies above the intercept, the demand exceeds the supply, forcing a response roll off and defining a bandwidth limit at f_{i2}. At this intercept frequency, the $1/\beta_2$ and A_{OL2} curves occupy the same point, making $1/\beta_2 = A_{OL2}$.

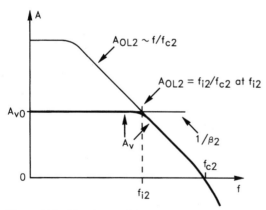

Figure 7.6 The A_2 amplifier of Fig. 7.5 introduces a second bandwidth limit defined by this amplifier's intercept of its $1/\beta$ curve and A_{OL} response at f_{i2}.

High-Gain Photodiode Amplifiers 145

This equality defines the f_{i2} bandwidth limit of A_2 when known quantities replace the above β_2 and A_{OL2} terms. First, the voltage amplifier formed with A_2 produces a feedback factor of $\beta_2 = R_1/(R_1+R_2)$, making $1/\beta_2 = 1 + R_2/R_1 = A_{v0}$. Next, the constant gain-bandwidth product of most op amps defines A_{OL2} at the intercept as guided by Fig. 7.6. There, A_{OL2} displays the typical single-pole roll off that produces this constant product over most of the amplifier's useful frequency range. The single-pole roll off produces a 1:1 slope for the A_{OL2} response, and an increase in closed-loop gain A_{v0} produces a proportionate decrease in the f_{i2} bandwidth limit. In this single-pole range, $A_{OL2} = f_{c2}/f$, where f_{c2} is the unity-gain crossover frequency of A_2. At the $1/\beta_2$ intercept, $A_{OL2} = A_{v0}$ and $f = f_{i2}$, making $A_{v0} = f_{c2}/f_{i2}$. Solving for f_{i2} defines the bandwidth limit for A_2 at $f_{i2} = f_{c2}/A_{v0}$. Then, equating the two amplifier bandwidth limits, $f_{i2} = f_{pf1}$, and solving for R_f' defines the value of this resistance for the bandwidth-versus-noise optimum. For a given value of net transresistance R_T, the optimum R_f' is

$$R_f' = \sqrt{R_T/2\pi f_{c2} C_S}$$

After solving for R_f' above, the closed-loop gain setting of the second amplifier becomes $A_{v0o} = R_T/R_f'$.

Two performance changes, produced by the added gain, control the resulting bandwidth improvement. First, the added gain reduces the converter's feedback resistance from R_f to $R_f' = R_T/A_{v0o}$. For comparison, consider the same circuit gain for the basic and two-amplifier solutions. Then, $R_T = R_f$, making $R_f' = R_f/A_{v0o}$. This replaces the original bandwidth limit of $f_{pf} = 1/2\pi R_f C_S$ with $f_{pf1} = 1/2\pi R_f' C_S = A_{v0o}/2\pi R_f C_S$, increasing this bandwidth limit by a factor equal to A_{v0o}. However, this difference in the f_{pf} limit does not improve the circuit's bandwidth by exactly the same factor due to a second effect. The added gain of the circuit also replaces the normal single-pole roll off with that of a double pole. This double pole results from the optimization described above, which increases A_{v0} until the two bandwidth limits coincide with $f_{pf1} = f_{i2}$. The two-pole roll off makes the circuit's -3-dB bandwidth $BW_t = 0.64 f_{pf1} = 0.64/2\pi R_f' C_S$, reducing the bandwidth improvement factor from A_{v0o} to $0.64\,A_{v0o}$.

Further manipulation of preceding results produces a simple indicator of the potential bandwidth improvement for a given application. Numerical evaluation of this indicator identifies those specific applications that would benefit from conversion to the two-amplifier solution. First, substituting the previous $R_f' = \sqrt{R_T/2\pi f_{c2} C_S}$ in the $A_{v0o} = R_T/R_f'$ result produces an alternate design equation

$$A_{v0o} = \sqrt{2\pi R_T C_S f_{c2}}$$

This expression defines A_{v0o} directly without first determining R_f'. More importantly, the variables of this expression also relate A_{v0o} to the original bandwidth limit f_{pf} and the crossover frequency f_{c2} of the added amplifier. From before, the basic circuit produces $f_{pf} = 1/2\pi R_f C_S$ and, for the same net gain, the circuit here makes $R_T = R_f$. Thus, for comparison purposes, $f_{pf} = 1/2\pi R_T C_S$ and $2\pi R_T C_S = 1/f_{pf}$. Substitution then yields

$$A_{v0o} = \sqrt{\frac{f_{c2}}{f_{pf}}}$$

This defines the optimum A_{v0} value in terms of the original bandwidth limit and the crossover frequency of A_2. Then, the preceding bandwidth improvement factor of 0.64 A_{v0o} defines the improvement available for a given application. Bandwidth improves with the two-amplifier solution whenever the above equation yields $A_{v0o} > 1/0.64 = 1.56$.

7.2.3 Noise analysis for the two-amplifier case

In optimizing the bandwidth-versus-noise compromise, the addition of voltage gain modifies each of the photodiode amplifier's fundamental noise components E_{noR}, E_{noi}, and E_{noe}. A listing at the end of this section summarizes these modified components and various subcomponents for step-by-step calculation of a given circuit's output noise. The three fundamental components represent the circuit's rms output noise created by resistance, noise current, and noise voltage as originally analyzed in Chap. 5. Here, the resistance and current noise effects, E_{noR} and E_{noi}, increase due to the added voltage gain and the increased bandwidth. However, they also decrease due to the reduction in R_f and this circuit's two-pole roll off. The added gain and steeper roll off also affect the circuit's response to the amplifier's input noise voltage E_{noe} in a similar manner. In addition, the presence of the second amplifier here introduces a new contributor to E_{noe}.

For the resistance and current noise components, E_{noR} and E_{noi}, the high-value feedback resistor R_f' continues to dominate related noise effects. The noise sources behind E_{noR} and E_{noi} remain the same, but the two-amplifier solution alters both the noise gain and its frequency response. By itself, resistor R_f' produces a constant noise density of $e_{nR} = \sqrt{4KTR_f'}$ that in turn produces the rms output noise component E_{noR}. Similarly, the amplifier's input bias current I_{B-} produces component E_{noi} through a constant noise density of $i_{ni} = \sqrt{2qI_{B-}}$. This noise current flows through R_f', developing a noise voltage of density

$e_{nRi} = i_{ni}R_f'$. Previously, the basic current-to-voltage converter transferred these resistance and current noise densities to the circuit output with unity gain at lower frequencies. Then, at higher frequencies, the single-pole bandwidth limit, BW_t of the circuit's transresistance, rolled off this gain. This combination made the corresponding noise gain $A_n = 1/(1 + jf/BW_t)$, where $BW_t = f_{pf} = 1/2\pi R_f C_S$.

In this two-amplifier case, the circuit instead transfers these noise densities to the circuit output with a gain of A_{v0} at lower frequencies. Then, at higher frequencies, a two-pole bandwidth limit rolls off A_{v0} at f_{pf1}, making $A_n = A_{v0}/(1 + jf/f_{pf1})^2$. Using this modified A_n result, evaluation of the rms noise integral

$$E_{no}^2 = \int_0^\infty |A_n e_n|^2 \, df$$

defines the E_{noR} and E_{noi} components for the two-amplifier case. In the first case, e_n of the above integral becomes e_{nR} and, in the second case, $i_{ni}R_f'$. For the two-amplifier solution of Fig. 7.4 this integration produces

$$E_{noR} = A_{v0}\sqrt{KTR_f'\pi f_{pf1}} \quad \text{and} \quad E_{noi} = A_{v0}R_f'\sqrt{q\pi f_{pf1} I_{B-}/2}$$

where $f_{pf1} = 1/2\pi R_f' C_S$, and $A_{v0} = 1 + R_2/R_1$.

Comparing these new noise results with earlier ones reveals the relative increases in E_{noR} and E_{noi} produced by this two-amplifier solution. For this comparison, two translations produce equivalent conditions. The first translates the resistance and bandwidth factors of the two-amplifier E_{noR} and E_{noi} results above. The A_{v0} gain added here modifies the circuit's conversion resistance from R_f to R_f' and the bandwidth limit from f_{pf} to f_{pf1}. Relating these modified characteristics to the originals produces a common ground for comparison of noise results. For a given transresistance, $R_T = R_f$, but the two-amplifier solution also makes $R_T = A_{v0}R_f'$. Solving for R_f' relates it to R_f through $R_f' = R_f/A_{v0}$. The two-amplifier solution also increases the fundamental f_{pf} bandwidth limit to $f_{pf1} = A_{v0}f_{pf}$. Substituting these R_f' and f_{pf1} expressions in the above E_{noR} and E_{noi} equations produces $E_{noR} = A_{v0}\sqrt{KTR_f\pi f_{pf}}$ and $E_{noi} = R_f\sqrt{q\pi A_{v0}f_{pf}I_{B-}/2}$. This expresses the modified E_{noR} and E_{noi} results in terms of the fundamental R_f and f_{pf} characteristics. A second translation converts the results of the basic circuit to the same common-ground form. Previously, analysis of the basic current-to-voltage converter produced $E_{noR} = \sqrt{2KTR_f\pi BW_t}$ and $E_{noi} = R_f\sqrt{q\pi BW_t I_{B-}}$. There, $BW_t = f_{pf}$, making $E_{noR} = \sqrt{2KTR_f\pi f_{pf}}$ and $E_{noi} = R_f\sqrt{q\pi f_{pf} I_{B-}}$ the equivalent noise results for comparison purposes. Comparing the translated results for the two-amplifier and basic circuits shows that the two-amplifier solution increases E_{noR} by a factor of $A_{v0}/\sqrt{2}$ and E_{noi} by a factor of $\sqrt{A_{v0}/2}$.

Amplifier noise voltages produce the third fundamental noise component, E_{noe}, but with the effects of two amplifiers in this case. Amplifier A_1 produces the noise gain peaking analyzed in Sec. 5.2.3, and then, A_2 amplifies the gain peaking result. Further, amplifier A_2 adds its own noise voltage to the net output noise result. Adding the subscripts 1 and 2 differentiates the A_1 and A_2 origins of the various noise sources and frequency limits described below. Consider the gain peaking effect first as represented by the spectral noise density versus frequency curve of Fig. 7.7. As before, separation of the spectral output noise e_{noe} into approximation regions simplifies the rms noise integration. Separate analysis of each region produces the rms noise components E_{noe1} through E_{noe5} for later combination into a net E_{noe} result. The e_{noe} curve shown incorporates three changes to the original curve of Fig. 5.5. The curve's magnitude increases, the span of region 2 expands, and the region 4 plateau vanishes. First, adding the voltage gain of A_2 raises the magnitude of the e_{noe} curve by a factor of A_{v0}, as reflected by the increased noise floor level $A_{v0}e_{nif1}$. Next, making the circuit's two bandwidth limits coincident, $f_{pf1} = f_{i2}$, extends the span

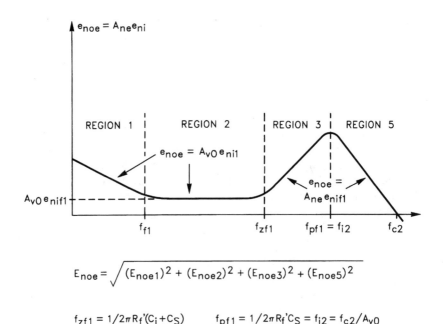

Figure 7.7 Amplifier A_1 of Fig. 7.4 produces an output noise density elevated by the gain A_{v0} of A_2 and missing the old region 4 plateau. Making the circuit's two bandwidth limits coincident removes that plateau.

of region 2 by moving f_{zf1} and f_{pf1} to the right. Substitution of f_{zf1} and f_{pf1} for the old f_{zf} and f_{pf} adjusts for this difference. Finally, the $f_{pf1} = f_{i2}$ alignment eliminates the old region 4 plateau by placing the plateau's beginning and ending points at the same frequency, making $E_{noe4} = 0$.

In addition to the above changes, adding A_2 inserts this amplifier's noise into the final result. The input noise voltage of this amplifier, e_{ni2}, receives the same A_v gain as the output signal of the circuit's A_1 current-to-voltage converter. Noise voltage e_{ni2} produces the typical $1/f$ response region, and this would complicate the associated analysis. However, the limited frequency range of this $1/f$ region and the signal preamplification provided by A_1 almost always reduce this region's added effect to insignificance. For simplicity, consider the input noise voltage of A_2 as just the amplifier's noise floor level e_{nif2}. Then, this constant noise density receives the gain, $A_v = A_{v0}/(1 + jf/f_{i2})$. Together, the A_2 noise voltage and voltage gain produce a new component of output noise, E_{noA2}. RMS evaluation of this component results from the integral

$$E_{noA2}^2 = \int_0^\infty |A_v e_{nif2}|^2 \, df$$

This produces $E_{noA2} = A_{v0} e_{nif2} \sqrt{\pi f_{i2}/2}$, where $A_{v0} = (1 + R_2/R_1)$, $f_{i2} = f_{c2}/A_{v0}$, and f_{c2} is the unity-gain crossover frequency of A_2. Combining E_{noA2} with the E_{noA1} result of the current-to-voltage converter, in rms fashion, then yields the circuit's final rms output noise.

The listing below summarizes the noise analysis of the two-amplifier solution of Fig. 7.4 for step-by-step calculation of the individual noise components and the resulting total output noise. For simplification, the listing defines A_{v0} and the various limit frequencies of the results only upon their first occurrences.

$E_{noe1} = A_{v0} e_{nif1} \sqrt{f_{f1} \ln(f_{f1}/f_1)}$, where $A_{v0} = 1 + R_2/R_1$. Also, e_{nif1} is the e_{ni1} noise floor level, and f_{f1} is the $1/f$ noise corner frequency of A_1. As explained in Sec. 5.2.3, setting $f_1 \approx 0.01$ Hz avoids a theoretical contradiction and ensures analytical accuracy.

$E_{noe2} = A_{v0} e_{nif1} \sqrt{f_{zf1} - f_{f1}}$, where $f_{zf1} = 1/2\pi R_f'(C_i + C_S)$, and $C_i = C_D + C_{id} + C_{icm}$. For applications requiring feedback compensation C_f, replace C_S with $C_f + C_S$ here and in the steps that follow. Numerical evaluation here can produce the square root of a negative number. Then, substitute zero for the result as explained in Sec. 5.3.1.

$E_{\text{noe3}} = A_{v0}(e_{\text{nif1}}/f_{\text{zf1}})\sqrt{\left(f_{\text{pf1}}^3 - f_{\text{zf1}}^3\right)/3}$, where $f_{\text{pf1}} = 1/2\pi R_f{'} C_S$, and C_S is the stray feedback capacitance.

$E_{\text{noe4}} = 0$.

$E_{\text{noe5}} = A_{v0}(e_{\text{nif1}}f_{c2})\sqrt{1/f_{i2}}$, where f_{c2} is the unity-gain crossover frequency of A_2, and $f_{i2} = f_{\text{pf1}} = 1/2\pi R_f{'} C_S$.

$E_{\text{noe}} = \sqrt{E_{\text{noe1}}^2 + E_{\text{noe2}}^2 + E_{\text{noe3}}^2 + E_{\text{noe5}}^2}$, the total rms output noise resulting from the spectral noise voltage e_{nil} of A_1.

$E_{\text{noR}} = A_{v0}\sqrt{KTR_f{'}\pi f_{\text{pf1}}}$, where $f_{\text{pf1}} = 1/2\pi R_f{'} C_S$.

$E_{\text{noi}} = A_{v0}R_f{'}\sqrt{q\pi f_{\text{pf1}} I_{B-}/2}$.

$E_{\text{noA1}} = \sqrt{E_{\text{noR}}^2 + E_{\text{noi}}^2 + E_{\text{noe}}^2}$, the total output noise due to the A_1 current-to-voltage converter.

$E_{\text{noA2}} = A_{v0}e_{\text{nif2}}\sqrt{\pi f_{i2}/2}$, the output noise due to the A_2 voltage amplifier. Here, $f_{i2} = f_{c2}/A_{v0}$ and f_{c2} is the unity-gain crossover frequency of A_2.

$E_{\text{no}} = \sqrt{E_{\text{noA1}}^2 + E_{\text{noA2}}^2}$, the total rms output noise of the two-op-amp photodiode amplifier.

7.3 Adding Voltage Gain

For many photodiode applications, the two op amps per photodetector above represents a significant drawback. Large arrays of photodetectors sometimes require hundreds of photodiode amplifiers. For such cases, one op amp provides both the current-to-voltage conversion and the voltage gain with somewhat less bandwidth improvement. The obvious circuit for this solution introduces nonlinearity and degrades rather than improves bandwidth. However, a bootstrap modification to this circuit avoids the nonlinearity and restores much of the bandwidth improvement of the two-amplifier solution.

7.3.1 Voltage gain alternatives

Evaluating the obvious circuit first illustrates the limitations later removed by the bootstrap. To provide both current-to-voltage conversion and voltage gain, this circuit first moves the conversion resistance $R_f{'}$ to the noninverting input of the op amp as in Fig. 7.8. There, the signal current supplied by D_1 still flows through $R_f{'}$ and again produces the desired signal voltage $i_p R_f{'}$. Moving $R_f{'}$ also makes available the amplifier's normal feedback path for the addition of voltage gain. There, the addition of R_1 and R_2 sets this gain, converting the op amp to

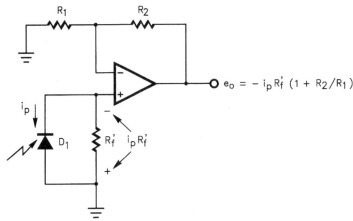

Figure 7.8 Using one op amp, the straightforward combination of current-to-voltage conversion and voltage gain impresses signal voltage on the diode, producing nonlinearity and restricting bandwidth.

a noninverting voltage amplifier. This amplifier provides the voltage gain $(1+R_2/R_1)$ to the $i_p R_f'$ signal developed across R_f'. However, this signal voltage also falls across the photodiode, reintroducing two performance limitations described in Chap. 2. This voltage-mode photodiode operation produces response nonlinearity, and the signal voltage develops a current in the photodiode capacitance that shunts the photodiode output. The shunting bypass of R_f' produces a response pole at $1/2\pi R_f' C_D$ and, for most applications, dramatically reduces bandwidth when compared with the basic current-to-voltage converter.

Avoiding these limitations, Fig. 7.9 bootstraps the photodiode by connecting it directly between the op amp's inputs. This connection removes the primary signal swing $i_p R_f'$ from the diode, restoring the linear response and most of the bandwidth improvement. Here, the circuit's resistors perform generally the same functions as in the preceding circuit. Current from the photodiode still flows through R_f', developing the same signal voltage, but that voltage no longer appears across the photodiode. The op amp feedback action reduces the voltage between the amplifier's inputs, and across the photodiode, to near zero. The bootstrap connection also supplies the diode current to the R_1, R_2 feedback network, altering the signal gain. However, the low resistances of this network produce little signal voltage in comparison with that produced by R_f'. Ignoring this added effect, the circuit again amplifies the voltage developed upon R_f', increasing the output signal to $e_o = (1+R_2/R_1)i_p R_f' = A_{vo} i_p R_f' = i_p R_T$. Here, $R_T = (1+R_2/R_1)R_f'$ represents the equivalent transresistance of the circuit.

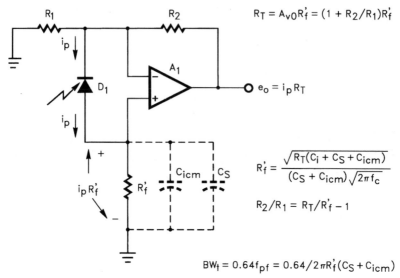

Figure 7.9 Bootstrapping the photodiode between the op amp inputs removes signal swing from the diode while retaining the signal gain of Fig. 7.8.

7.3.2 Optimizing the single-amplifier bandwidth-versus-noise compromise

This single-amplifier connection develops the same net transresistance as the preceding two-amplifier solution. However, the bandwidth improvement realized decreases here. A modified bandwidth limit results due to the move of the high-value conversion resistance R_f'. That resistor now connects to the noninverting input of the op amp where the common-mode input capacitance of the op amp C_{icm} also shunts R_f'. Previously, only the smaller stray capacitance, C_S of the two-amplifier circuit, shunted this resistor in its feedback connection. The increased shunting here rolls off bandwidth at $f_{pf} = 1/2\pi R_f'(C_S + C_{icm})$ rather than the preceding $f_{pf1} = 1/2\pi R_f' C_S$. Thus, compared with the two-amplifier case, the single-amplifier solution degrades the bandwidth improvement by the factor $1 + C_{icm}/C_S$.

For a given transresistance, $R_T = A_{v0}R_f'$, and increasing A_{v0} again reduces the required R_f', moving this modified bandwidth limit to a higher frequency. However, increasing A_{v0} also reduces the frequency of the separate bandwidth limit imposed by the op amp. The amplifier imposes a bandwidth limit again marked by its $1/\beta$ intercept with the A_{OL} response. As described with the two-amplifier analyses, this limit occurs at $f_i = \beta f_c$ where β and f_c are the feedback factor and the unity-gain crossover frequency of the op amp. Increasing A_{v0} decreases β, reducing the frequency of the f_i bandwidth limit. In compromise, increasing A_{v0} until the circuit's two bandwidth limits coincide, $f_{\text{pf}} = f_i = \beta f_c$, maximizes the final circuit bandwidth with neither limit imposing a dominant constraint. This choice also optimizes the circuit's noise-versus-bandwidth compromise as detailed in the discussion of Sec. 7.2.1.

Analysis quantifies the potential bandwidth improvement of the single-amplifier solution through comparison with the basic current-to-voltage converter. Adding voltage gain to reduce the basic R_f to R_f' initially increases bandwidth by reducing the sensitivity to parasitic capacitance. However, two effects described above limit this increase in the single-amplifier case. First, setting $f_i = f_{\text{pf}}$ produces a double rather than single-pole roll off for signal bandwidth, just as described for the two-amplifier case. This reduces signal bandwidth from $\text{BW}_t = f_{\text{pf}}$ to $\text{BW}_t = 0.64 f_{\text{pf}}$. Next, the added shunting of the amplifier's input capacitance tends to reduce the f_{pf} pole frequency, as seen by comparing the basic $f_{\text{pf}} = 1/2\pi R_f C_S$ with the new $f_{\text{pf}} = 1/2\pi R_f'(C_S + C_{\text{icm}})$. Noting that $R_f' = R_T/A_{v0}$, the new bandwidth limit becomes $\text{BW}_t = 0.64 A_{v0}/2\pi R_T(C_S + C_{\text{icm}})$, where R_T represents the net transresistance of the circuit. For the same transresistance, the current-to-voltage converter alone requires a feedback resistance of $R_f = R_T$, making its $\text{BW}_t = 1/2\pi R_T C_S$ for comparison purposes. Dividing this BW_t expression into the preceding one shows that bandwidth changes by a factor of $0.64 A_{v0}/(1 + C_{\text{icm}}/C_S)$. In cases where $A_{v0} > 1.56(1 + C_{\text{icm}}/C_S)$, the single-amplifier solution improves bandwidth over that of the basic current-to-voltage converter.

To quantify A_{v0} the preceding $f_{\text{pf}} = f_i = \beta f_c$ optimum first requires defining the circuit's feedback factor β. However, moving the photodiode to its bootstrap position complicates this task. Positioned between the amplifier inputs, the diode provides an added, positive feedback path between the amplifier's output and noninverting input. Figure 7.10 models this connection for feedback analysis with $C_i = C_D + C_{\text{id}}$ combining the diode capacitance with the differential-input capacitance of the op amp. Through the feedback network, C_i couples a portion of output signal e_o to R_f' and the amplifier's noninverting input. This added coupling results in one feedback network that supplies feedback signals to both op amp inputs.

Normally, separate analyses of negative and positive feedback factors, β_- and β_+, yields individual results for determination of the net feedback factor $\beta = \beta_- - \beta_+$. However, the single feedback network here simplifies the analysis given a reexamination of the feedback factor definition. Fundamentally, feedback factor is the fraction of the output signal fed back to the amplifier input. With two inputs, an op amp departs somewhat from this fundamental definition. The

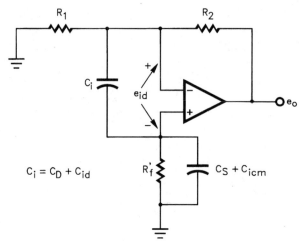

Figure 7.10 The photodiode bootstrap of Fig. 7.9 adds capacitance between the op amp inputs, coupling a positive feedback signal to R_f. One feedback factor analysis combines the circuit's negative and positive feedback effects by solving for $\beta = e_{\text{id}}/e_o$.

differential nature of the two inputs subtracts the feedback signals supplied to the two, resulting in a net differential feedback signal e_{id}. Then, $e_{\text{id}}/e_o = \beta$, and one analysis combines the separate β_- and β_+ analyses that otherwise produce more complex intermediate results. In this analysis, the approximation $R_1 \ll Z_f'$ further simplifies the result, where Z_f' represents the parallel impedance combination of $C_S + C_{\text{icm}}$ and R_f'. The low value of $C_S + C_{\text{icm}}$ and the high value of R_f' support this approximation for high-gain photodiode amplifiers. Then, analyzing Fig. 7.10 for $e_{\text{id}}/e_o = \beta$ produces

$$\beta = \beta_0 \frac{1 + R_f'(C_S + C_{\text{icm}})s}{1 + R_f'(C_i + C_S + C_{\text{icm}})s}$$

where $\beta_0 = R_1/(R_1 + R_2) = 1/A_{v0} = R_f'/R_T$.

Manipulation of this β result and application of the $f_{\text{pf}} = f_i = \beta f_c$ condition defines R_f' and A_{v0} for the optimum. First, $f_i = \beta f_c$ defines the frequency of the $1/\beta$ intercept, and this occurs at a higher frequency where β equals its high-frequency value, β_H. There, the circuit's capacitances control the feedback and the preceding β expression reduces to

$$\beta_H = \frac{\beta_0(C_S + C_{\text{icm}})}{C_i + C_S + C_{\text{icm}}}$$

High-Gain Photodiode Amplifiers   155

Substituting $\beta_0 = R_f'/R_T$, reduces this to

$$\beta_H = \frac{R_f'}{R_T} \frac{C_S + C_{\text{icm}}}{C_i + C_S + C_{\text{icm}}}$$

Then, equating $f_{\text{pf}} = 1/2\pi R_f'(C_S + C_{\text{icm}})$ and $f_i = \beta_H f_c$ yields the key design equation for the single-amplifier solution

$$R_f' = \frac{\sqrt{R_T(C_i + C_S + C_{\text{icm}})}}{(C_S + C_{\text{icm}})\sqrt{\omega_c}}$$

where $\omega_c = 1/2\pi f_c$.

Application of this equation defines the optimum A_{v0} and circuit resistances for a given application. First, the specific application defines the value of R_T required to produce the desired transresistance. Then, the equation defines the value of R_f' that optimizes the circuit's bandwidth-versus-noise compromise. Given R_T and R_f', the relationship $A_{v0} = R_T/R_f'$ defines A_{v0}. Next, the relationship $A_{v0} = 1 + R_2/R_1$ defines the ratio $R_2/R_1 = R_T/R_f' - 1$. A second consideration converts this ratio into the final values for R_1 and R_2. Together, these resistors present a load resistance of $R_L = R_1 + R_2$ to the op amp output. Picking a convenient resistance value for this load defines the individual resistors as $R_1 = (R_f'/R_T)R_L$ and $R_2 = (1 - R_f'/R_T)R_L$. These design choices for R_f', R_1, and R_2 produce a single-amplifier combination of current-to-voltage conversion and voltage gain with optimum bandwidth and minimized noise increase.

7.3.3 Noise analysis for the single-amplifier case

The noise performance of the single-amplifier solution of Fig. 7.8 repeats that of the preceding two-amplifier case with only a few changes. The added voltage gain again amplifies all noise components, and the $f_i = f_{\text{pf}}$ setting results in an e_{noe} curve identical in shape to the earlier Fig. 7.7. Adding the effects of the C_{icm} bypass adapts the earlier results to this case.

At first, it might seem that the bootstrap circuit removes the gain peaking effect upon the amplifier input noise voltage. Previously, this noise voltage produced a noise current with the diode's capacitance, and that current flowed through the large feedback resistance, producing the noise gain peaking. The bootstrap circuit here removes the large resistance from the feedback path but replaces it with the resistor R_f' at the amplifier's noninverting input. Noise current generated in C_i, by the noise voltage between the op amp inputs, still flows in the reconnected R_f', producing much the same noise peaking effect. To quantify this effect, consider the feedback factor expression β developed above. The op amp amplifies its input noise voltage by a gain of $1/\beta$ up to the gain roll off imposed by the amplifier's A_{OL} decline. From the previous β analysis, the bootstrap photodiode amplifier produces the noise gain

$$1/\beta = A_{v0} \frac{1 + R_f'(C_i + C_S + C_{icm})s}{1 + R_f'(C_S + C_{icm})s}$$

where $A_{v0} = 1/\beta_0 = 1 + R_2/R_1$. Thus, the noise gain displays the response zero, $f_{zf} = 1/2\pi R_f'(C_i + C_S + C_{icm})$, and the response pole, $f_{pf} = 1/2\pi R_f'(C_S + C_{icm})$, that characterize noise gain peaking. Together, the added voltage gain A_{v0} and the gain peaking effect reproduce the Fig. 7.7 e_{noe} curve with just the f_{zf} and f_{pf} differences here noted in the listing below. Then, calculation of the noise result for the Fig. 7.9 bootstrap amplifier follows the step-by-step procedure:

$E_{noe1} = A_{v0} e_{nif} \sqrt{f_f \ln(f_f/f_1)}$, where $A_{v0} = 1 + R_2/R_1$. Also, e_{nif} is the e_{ni} noise floor level, and f_f is the $1/f$ noise corner frequency of the op amp. As explained in Sec. 5.2.3, setting $f_1 \approx 0.01$ Hz avoids a theoretical contradiction and ensures analytical accuracy.

$E_{noe2} = A_{v0} e_{nif} \sqrt{f_{zf} - f_f}$, where $f_{zf} = 1/2\pi R_f'(C_i + C_S + C_{icm})$, where $C_i = C_D + C_{id}$, C_S is the stray capacitance, and C_{icm} is the common-mode input capacitance of the op amp. Numerical evaluation here can produce the square root of a negative number. Then, substitute zero for the result as explained in Sec. 5.3.1.

$E_{noe3} = A_{v0}(e_{nif}/f_{zf})\sqrt{\left(f_{pf}^3 - f_{zf}^3\right)/3}$, where $f_{pf} = 1/2\pi R_f'(C_S + C_{icm})$.

$E_{noe4} = 0$.

$E_{noe5} = A_{v0}(e_{nif} f_{c2})\sqrt{1/f_i}$, where f_c is the unity-gain crossover frequency of the amplifier.

$E_{noe} = \sqrt{E_{noe1}^2 + E_{noe2}^2 + E_{noe3}^2 + E_{noe5}^2}$, the total rms output noise resulting from the spectral noise voltage e_{ni} of the amplifier input.

$E_{noR} = A_{v0}\sqrt{KTR_f' \pi f_{pf}}$.

$E_{noi} = A_{v0} R_f' \sqrt{q \pi f_{pf} I_{B-}/2}$.

$E_{no} = \sqrt{E_{noR}^2 + E_{noi}^2 + E_{noe}^2}$, the total rms output noise of the bootstrap photodiode amplifier.

Compared with the two-amplifier case, the above results eliminate E_{noeA2} and display changes in C_i, f_{zf}, and f_{pf}. Noise voltage E_{noeA2}, the noise of A_2 in the previous results, no longer exists for this single-amplifier solution. Capacitance C_i becomes $C_i = C_D + C_{id}$, in this case, eliminating the previous C_{icm} component. For this circuit, the role of C_{icm} transfers to the bypass R_f's signal voltage. This bypass alters f_{zf} and f_{pf} as described in the $1/\beta$ analysis where $f_{zf} = 1/2\pi R_f'(C_i + C_S + C_{icm})$ and $f_{pf} = 1/2\pi R_f'(C_S + C_{icm})$.

7.4 Adding Current Gain

Adding current rather than voltage gain presents another bandwidth-improving alternative to high-gain photodiode amplifiers. This alternative best serves applications requiring a current output and, there, current gain replaces transimpedance as the amplifier's gain mechanism. However, for photodiode sources, direct application of the op amp current amplifier produces nonlinearity, latching, and oscillations. Like the single-amplifier case above, bootstrapping the photodiode avoids these problems by removing signal swing from the photodiode. Then, judicious component selection again optimizes the circuit's noise-versus-bandwidth compromise. A simple feedback analysis transforms previous noise results to this current-gain case.

7.4.1 Current gain alternatives

Op amp circuits readily produce current rather than voltage gain,[2] and direct application of the basic, op amp current amplifier to photodiodes produces Fig. 7.11. Analysis of this circuit's performance first defines

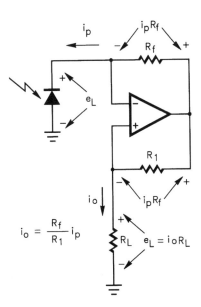

Figure 7.11 Connecting a photodiode directly to an op-amp current amplifier adds current gain but impresses signal voltage e_L across the diode, producing nonlinearity and latching.

the problems of the basic current-gain case, and then analysis of the bootstrap alternative illustrates the solution. In the figure, negative and positive feedback combine to deliver a current gain controlled by two resistors, R_f and R_1. This circuit actually performs a current-to-voltage conversion with R_f followed by the inverse function with R_1, adding gain in the process. The added gain again permits reducing R_f to increase bandwidth, and the circuit's single-amplifier structure avoids the noise of a second amplifier. Further, the circuit's dual function provides an extra degree of design freedom that again permits optimization of the noise-versus-bandwidth compromise.

Temporarily neglecting capacitance effects illustrates this dual function to define the circuit's low-frequency current gain A_{i0}. At lower frequencies, signal current i_p from the photodiode flows through R_f, producing the input to output voltage drop $i_p R_f$. Feedback replicates this voltage across R_1 to retain zero voltage between the op amp inputs. In doing this, the feedback action supplies the current $i_o = i_p R_f / R_1$ through R_1 and to the load R_L. Thus, the circuit's current gain i_o/i_p is $A_{i0} = R_f/R_1$ at low frequencies. This added gain permits the bandwidth-improving reduction of R_f, as described for the two preceding circuits.

However, this direct application of the basic current amplifier produces three severe problems for the photodiode amplifier.[3] Each of these problems results from impressing the load voltage across the photodiode. In the first, nonlinearity develops from the inherent variation of photodiode responsivity with the diode's voltage. In Fig. 7.11, feedback replicates the load's signal voltage e_L directly across the diode, modulating the diode's responsivity. In a second problem, this signal condition produces a latch state when a larger e_L causes forward bias of the photodiode. Normally, the output signal current i_o develops an e_L voltage polarity that holds the diode in a reverse-biased state. However, the transient condition of circuit turn on can momentarily produce a forward bias. Then, the diode shorts the negative feedback of R_f to ground, leaving the positive feedback of R_1 in control of the circuit. This positive feedback dominance latches the circuit in this forward-biased state. Limiting the load impedance R_L reduces the possibility of the larger e_L required to produce this latch state. However, the uncertainty of transient conditions during circuit turn on make the circuit unreliable.

In a third problem condition, impressing e_L across the photodiode capacitance promotes oscillation as illustrated by Fig. 7.12. Voltage e_L results from a positive feedback connection and its drop across capacitance C_D produces a positive feedback current. This current flows to R_f, transferring a positive feedback signal to the negative feedback path, inviting oscillation. From another perspective, capacitance

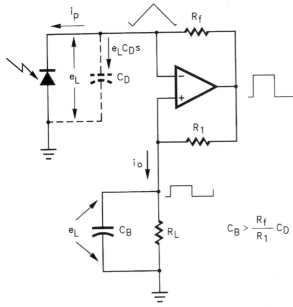

Figure 7.12 The e_L drop across the photodiode capacitance also promotes oscillation with the Fig. 7.11 connection.

C_D rolls off the negative feedback of R_f, making it possible for the circuit's positive feedback to dominate through R_1. In fact, this feedback combination inadvertently converts the circuit to the basic op amp square-wave generator.[4] For this generator, the op amp acts as a comparator, producing a square wave output in response to triangle- and square-wave signals at the amplifier's two inputs. These input signals result from an integrating negative feedback and a hysteresis-producing positive feedback. Here, R_f and C_D integrate the square-wave output signal, producing a triangle wave at the amplifier's inverting input. Meanwhile, R_1 and R_L transfer a square wave to the amplifier's noninverting input, shifting the threshold for the ensuing comparator action of the amplifier. Each time the triangle wave at the inverting input reaches the level of the square wave at the noninverting input, the circuit switches comparator states. To prevent this oscillation, a dominant roll off must be added to the positive feedback path through the load bypass of C_B. This requires a bypass capacitance $C_B > A_{io}C_D = R_fC_D/R_1$ and devastates signal bandwidth.

Bootstrapping the photodiode avoids all three of the above problems by removing the load's signal voltage from the photodiode. Figure 7.13 shows this solution with the previously grounded end of the diode returned to the top of load R_L. Then, signal voltage e_L no

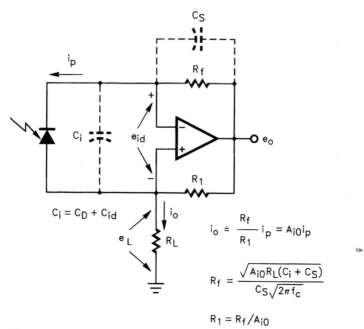

Figure 7.13 Bootstrapping the photodiode removes e_L from the diode, avoiding the problems of Fig. 7.11 and delivering the desired current gain $A_{i0} = i_o/i_p = R_f/R_1$.

longer drops across the diode, avoiding the nonlinearity, latching, and oscillation of the preceding circuit. Here, the e_L signal drives the op amp's noninverting input, and feedback virtually duplicates this signal at the amplifier's inverting input. Thus, the positive feedback signal becomes a common-mode signal at the amplifier's two inputs, and common-mode rejection virtually removes the signal from the diode. Only the small input error signal of the op amp, e_{id}, remains across the photodiode. This very small signal leaves the diode responsivity essentially unchanged and avoids the forward-biased diode state that produced latching. Also, removing the signal voltage from the diode's capacitance avoids the coupling of a positive feedback current into the negative feedback path. The negative feedback network now rides atop the positive feedback signal, preventing interaction. Given these cures, the bootstrapped current amplifier reliably produces the current gain $A_{i0} = i_o/i_p = R_f/R_1$ as originally intended for the previous circuit. The bootstrap connection actually increases this gain slightly due to the added flow of i_p from the diode's anode directly to R_L. However, high-gain applications typically produce $i_o \gg i_p$, making this added gain negligible.

7.4.2 Optimizing the current-gain bandwidth-versus-noise compromise

This compromise parallels that of the two-amplifier solution previously described in Sec. 7.2.1. There and here, both the bandwidth and noise performance of the circuit exhibit peaks in their relationships with the added gain. The two peaks occur at the same gain value, and the optimum bandwidth-versus-noise performance occurs at this peak value. The earlier discussion defines the origins of this optimum. As before, a specific A_{i0} value here optimizes bandwidth and the bandwidth-versus-noise compromise simultaneously. Thus, an analysis focus upon bandwidth versus A_{i0} automatically includes the noise compromise as well.

The bootstrapped current amplifier returns bandwidth control to the limits normally imposed upon photodiode amplifiers. Bootstrap removes the positive feedback coupling that previously required the C_B bypass of the load in Fig. 7.12. That bypass dramatically reduced signal bandwidth as described before. Removing this bypass, the bootstrap alternative returns bandwidth control to the limits imposed by capacitive bypass of R_f and the A_{OL} roll off of the op amp. In high-gain cases, the parasitic C_S bypass of R_f sets a bandwidth limit at the pole frequency $f_{pf} = 1/2\pi R_f C_S$. As before, adding circuit gain permits reduction of R_f and increases this f_{pf} limit. However, the second bandwidth limit, produced by the amplifier's A_{OL} roll off, ultimately restricts the degree of improvement. This roll off results from the amplifier's limited gain-bandwidth product beyond which the amplifier lacks the loop gain required to support the circuit's ideal response, $A_{i0} = i_o/i_p$. The intercept of the circuit's $1/\beta$ curve and the A_{OL} response defines the maximum frequency for which A_{OL} supports the gain demand of the feedback. As described with the earlier Fig. 7.6, this intercept imposes a bandwidth limit at $f_i = \beta_H f_c$, where β_H is the high-frequency value of the circuit's feedback factor β, and f_c is the unity-gain crossover frequency of A_{OL}. Making the f_{pf} and f_i bandwidth limits coincident optimizes bandwidth and defines design equations for the bootstrapped current amplifier. Aligning f_i with f_{pf} again produces a two-pole roll off for the signal response, making $BW_t = 0.64 f_{pf}$, where $f_{pf} = 1/2\pi R_f C_S$.

Equating the expressions for f_{pf} and f_i defines optimum-bandwidth design equations for R_f and R_1 in terms of known circuit quantities. However, the combined negative and positive feedback paths complicate the determination of $f_i = \beta_H f_c$. This requires knowledge of the circuit's feedback factor β, and the two feedback paths both affect this factor. Still, the combined effect of the two develops a single differential input signal e_{id} between the op amp's inputs. This signal defines the fraction of the output fed back to the amplifier's differential input or the feedback factor. Analyzing the circuit for $e_{id}/e_o = \beta$ produces a complex result, but practical conditions reduce it to a simple equation. In practice, $Z_f \gg R_1$, where Z_f is the paralleled combination of R_f and C_S. The low value of C_S and, for high-gain cases, the high value of R_f make $Z_f \gg R_1$ over the useful range of the op amp bandwidth. Given this approximation, the circuit's feedback factor becomes

$$\beta = \frac{R_1}{R_1 + R_L} \frac{1 + R_f C_S s}{1 + R_f(C_i + C_S)s}$$

where $C_i = C_D + C_{id}$. Here, C_D is the photodiode capacitance, and C_{id} is the differential-input capacitance of the op amp. At high frequencies, the circuit capacitances dominate, reducing this feedback factor to

$$\beta_H = \left(\frac{R_1}{R_1 + R_L}\right)\left(\frac{C_S}{C_i + C_S}\right) = \beta_0 \frac{C_S}{C_i + C_S}$$

Given this expression, $f_i = \beta_H f_c$ defines f_i for the coincident bandwidth limit of the $f_{pf} = f_i$ bandwidth optimization. However, equating f_{pf} and f_i and solving for the required R_f yields another complex result. A practical consideration again offers a simplification through an approximation for β_0, the low-frequency value of β. Here, $\beta_0 = R_1/(R_1 + R_L)$, and output compliance considerations favor the condition $R_1 \ll R_L$, making $\beta_0 \approx R_1/R_L$. Output compliance range, or the voltage limit across R_L, results from the saturation voltage limit of the op amp output. For linear response, the amplifier output voltage e_o must support the voltage drops across both R_1 and R_L without encountering this saturation limit. These two resistances conduct the same current i_o, and making $R_1 \ll R_L$ reserves most of the amplifier's voltage range for the voltage across R_L. Then, $\beta_0 \approx R_1/R_L$, making $\beta_H = (R_1/R_L)C_S/(C_i + C_S)$ and $f_i = \beta_H f_c = (R_1/R_L)f_c C_S/(C_i + C_S)$. Equating this f_i expression to $f_{pf} = 1/2\pi R_f C_S$ and solving for R_f yields the design equation

$$R_f = \frac{\sqrt{A_{i0} R_L (C_i + C_S)}}{C_S \sqrt{\omega_c}}$$

where $C_i = C_D + C_{id}$ and $\omega_c = 1/2\pi f_c$. In this equation, conditions of the specific circuit define all of the variables except A_{i0}. This last variable depends upon the current gain requirement of the specific application, as chosen by the user. Choosing a value for A_{i0} for the requirement then defines R_f. Next, the design value for R_1 follows from $R_1 = R_f/A_{i0}$. Together, the design equations of this paragraph define the circuit conditions that optimize the circuit's gain-bandwidth compromise. These equations simultaneously optimize the circuit's bandwidth-versus-noise compromise as described earlier in this section.

7.4.3 Noise analysis for the current gain case

Noise analysis for the bootstrapped current amplifier requires two steps. First, the conventional E_{no} analysis defines the rms noise voltage at the amplifier output. Then, a conversion analysis relates the E_{no} result to the output noise current I_{no}. As before, E_{no} results from the noise components E_{noR}, E_{noi}, and E_{noe}. These components represent the rms output noise resulting from circuit resistance and from the input current and voltage noises of the amplifier. Gain and bandwidth differences distinguish the noise results here from those of the basic current-to-voltage converter. First, the circuit's positive feedback amplifies the basic noise effect for each noise component by a factor of $1/\beta_0$. Here, $\beta_0 = R_1/(R_1 + R_L)$ is the low-frequency value of the circuit's feedback factor as developed before. Next, the circuit's two-pole roll off, resulting from the $f_i = f_{pf}$ alignment, reduces the effect of each noise component through a steeper response roll off.

The amplified, two-pole results parallel those of the single-amplifier solution described in Sec. 7.3.2. Converting those earlier results to this case replaces the previous gain A_{v0} with $1/\beta_0$ here. Previously, the noise results reflected the performance of the similarly bootstrapped amplifier of Fig. 7.9. One other difference separates the two cases because of the change in the capacitance bypassing the R_f resistance. Here, stray capacitance C_S bypasses this resistance rather than the $C_S + C_{icm}$ before. Thus, replacing A_{v0} with $1/\beta_0$ and $C_S + C_{icm}$ with C_S translates the Sec. 7.3.3 noise results for application to this case. Then, for the Fig. 7.13 current amplifier the noise voltage at the amplifier output follows from the analysis steps below:

$E_{noe1} = (e_{nif}/\beta_0)\sqrt{f_f \ln(f_f/f_1)}$, where $1/\beta_0 = 1 + R_L/R_1$. Here, e_{nif} is the e_{ni} noise floor level, and f_f is the $1/f$ noise corner frequency of the op amp. As explained in Sec. 5.2.3, setting $f_1 \approx 0.01$ Hz avoids a theoretical contradiction and ensures analytical accuracy.

$E_{noe2} = (e_{nif}/\beta_0)\sqrt{f_{zf} - f_f}$, where $f_{zf} = 1/2\pi R_f(C_i + C_S)$, and $C_i = C_D + C_{id}$. Numerical evaluation here can produce the square root of a negative number. Then, substitute zero for the result as explained in Sec. 5.3.1.

$E_{noe3} = (e_{nif}/\beta_0 f_{zf})\sqrt{\left(f_{pf}^3 - f_{zf}^3\right)/3}$, where $f_{pf} = 1/2\pi R_f C_S$.

$E_{noe4} = 0$.

$E_{noe5} = (e_{nif} f_{c2}/\beta_0)\sqrt{1/f_i}$, where f_c is the unity-gain crossover frequency of the amplifier, and $f_i = f_{pf} = 1/2\pi R_f C_S$.

$E_{noe} = \sqrt{E_{noe1}^2 + E_{noe2}^2 + E_{noe3}^2 + E_{noe5}^2}$, the total rms output noise resulting from the spectral noise voltage e_{ni} of the amplifier input.

$E_{noR} = (1/\beta_0)\sqrt{KTR_f \pi f_{pf}}$

$E_{noi} = (R_f/\beta_0)\sqrt{q \pi f_{pf} I_{B-}/2}$

$E_{no} = \sqrt{E_{noR}^2 + E_{noi}^2 + E_{noe}^2}$, the total rms noise at the op amp output.

Having calculated E_{no}, the actual noise of interest I_{no} follows from a simple conversion. In Fig. 7.13, the positive feedback to the amplifier's noninverting input delivers a portion of e_o, including a portion of E_{no}, to the load R_L. This portion, or fraction of the output fed back to the noninverting input, equals the circuit's positive feedback factor β_+. Thus, the portion of E_{no} noise voltage delivered to R_L equals $(\beta_+)E_{no}$, making $I_{no} = (\beta_+)E_{no}/R_L$. Two feedback paths contribute to β_+ with the primary path through R_1. A secondary path develops through R_f, C_S, and C_i, but this path contributes much less feedback signal. For high-gain photodiode amplifiers, the parallel impedance of R_f and C_S greatly exceeds that of R_1 over the useful frequency range of the amplifier. Neglecting the secondary path, $\beta_+ = R_L/(R_1 + R_L)$, making the output noise current of Fig. 7.13

$$I_{no} = \frac{E_{no}}{R_1 + R_L}$$

References

1. J. Graeme, "The Tee-Feedback Factor in Photodiode Amplifiers," *Electronic Design*, June 26, 1995, p. 98.
2. J. Graeme, "Manipulate Current Signals with Op Amps," *EDN*, August 8, 1985, p. 147.
3. J. Graeme, "Tame Photodiodes with Op Amp Bootstrap," *Electronic Design*, May 12, 1988, p. 89.
4. G. Tobey, J. Graeme, and L. Huelsman, *Operational Amplifiers; Design and Applications*, McGraw-Hill, New York, 1971.

Chapter

8

Reducing Power-Supply Noise Coupling

Noise on the power-supply lines couples into the signal path of the photodiode amplifier through the finite PSRR of the op amp. This noise most often results from the line impedance interaction with currents drawn by the amplifier and other circuitry powered from the same supply lines. Supply current flows through these impedances, producing voltage drops that represent noise at the amplifier supply connections. This power-supply noise couples through the amplifier's PSRR to reproduce a portion of the noise signal at the op amp inputs. There, the coupled noise combines with the amplifier's normal input noise to be amplified by the noise gain peaking of the photodiode amplifier. This gain peaking effect makes supply noise reduction especially important for the photodiode amplifier. The gain peaking also makes the photodiode amplifier more vulnerable to oscillation. That portion of the supply noise resulting from the amplifier itself constitutes a parasitic feedback signal, capable of producing this self-generated noise condition.

Numerous techniques commonly reduce this parasitic coupling through wide supply buss traces, minimized buss lengths, star connections, and supply bypass. The first two techniques, buss width and length control, reduce supply-line resistance and inductance. Star connections provide separate supply runs to sensitive and noisy circuitry, such as analog and digital, to minimize device interaction. However, most importantly, capacitive bypass of the power-supply lines greatly attenuates the noise interference and ensures frequency stability as well. To be successful, the bypass selection requires close attention to the frequency-dependent impedances of both the supply lines and the bypass capacitors. Initial intuition suggests simply placing the bypass capacitors right at the amplifier's power-supply terminals and making the capacitors large.

Here, the first intuitive assumption works but the second fails. Close placement minimizes the interconnect length and the associated inductance between the amplifier and the capacitors. For best results, the bypass capacitors should be surface-mount types mounted right between the amplifier's power-supply pins and a ground plane. However, practical layout considerations often prevent this ideal condition, and in any case, amplifier lead impedance remains unbypassed. The second intuitive assumption, making the capacitors large, fails because of the parasitic impedances of such capacitors. Large capacitor values do minimize the bypassed line impedances at lower frequencies where the capacitors remain purely capacitive. However, the internal construction of larger capacitors produces a significant parasitic inductance, interrupting the bypass effectiveness at higher frequencies. In compromise, the primary bypass capacitance value should be made large enough to sufficiently limit line impedance but no larger.

Where necessary, adding a smaller capacitor in parallel with the first commonly bypasses the inductance of the first. However, the secondary capacitor has inductance of its own, producing another bypass compromise at a somewhat higher frequency. Further, the secondary capacitor reacts with the inductance of the first, producing a resonant increase in the supply-line impedance. Careful selection of these capacitors produces a compromise solution that retains low bypass impedance throughout the amplifier's significant frequency range. For this selection, analysis of the net bypass impedance produces two simple design equations that guide the capacitor selection. For the more severe coupling cases, a resistive detuning of the bypass impedance preserves stability, or the addition of series line impedance forms decoupling filters. For both of these special cases additional design equations again guide the component selection.

8.1 The Power-Supply Bypass Requirement

Noise on the power-supply lines of an op amp circuit couples to the circuit's output with a magnitude controlled by the supply-line impedance, the amplifier's PSRR, and the circuit's noise gain. Typically, the inductive portion of the line impedance produces the most serious coupling due to the associated impedance rise with increasing frequency. Simultaneous with this rise, the amplifier's PSRR declines, increasing the parasitic coupling further. For photodiode amplifiers, yet a third increase results from the characteristic noise gain peaking. Bypass capacitors convert the line impedance rise to a decline to restrain this coupling increase. In addition, these capacitors must pre-

Reducing Power-Supply Noise Coupling

serve frequency stability by sufficiently reducing the line impedance at higher frequencies. Part of the supply-line noise signal results from the amplifier itself and establishes a parasitic feedback loop capable of producing oscillation.

8.1.1 The noise-coupling mechanism

Figure 8.1 illustrates the fundamental need for power-supply bypass. There, positive and negative power supplies V_+ and V_- supply the corresponding bias terminals of the amplifier with the voltages V_P and V_N. Ideally, $V_P = V_+$ and $V_N = V_-$ but intervening, supply-line inductances L_p react with the signal current i_S, producing voltage differences. Here, the signal current i_S results from the combined supply currents drawn by the op amp and other circuitry powered from the same supply lines. The L_p inductances result from an inherent parasitic effect of the connecting lines between the power supplies and the amplifier. A typical printed circuit board trace introduces about 15 nH of parasitic inductance per inch of trace and wire introduces a similar value.[1] While seemingly small, complex board and wiring connections produce inductances in the hundreds of nanohenrys. The resulting supply-line impedances can seriously degrade noise performance and frequency stability. Adding the C_B bypass capacitors shown rolls off these impedances to attenuate the supply coupling effects.

The circuit model shown intentionally oversimplifies the actual conditions of a practical circuit to provide more intuitive insight into the bypass requirement. First, the model omits the input signal to permit

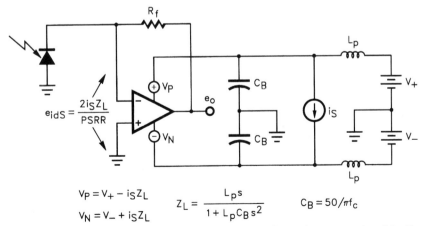

Figure 8.1 Bypass capacitors C_B suppress supply voltage changes produced by the reaction of supply current drain i_S with power-supply line inductances L_p.

focus upon only those signals relating to the line impedance and its bypass. Next, the model simplifies the line conditions. In practice, the V_+ and V_- supplies add output impedances and noise sources of their own, different circuits draw signal currents at different points upon the supply lines, and the supply-line resistance adds to the voltage drops produced with i_S. However, the bypass practices developed with this simplified model also contend with these other effects.

Intuitive evaluation of the model provides a first insight into the bypass benefit. Fundamentally, the C_B bypass capacitors shunt the line impedances to reduce the supply-line voltage drops produced by i_S. However, from another perspective, the capacitors serve as local reservoirs for the immediate supply of higher-frequency current demands. Otherwise, such currents encounter significant time delays in their travel from the power supplies and through the line inductances to the op amp. This delay produces phase shift in the amplifier response and, for a given time delay, the corresponding phase shift increases with frequency. To counteract this, the charge stored on the bypass capacitors supplies much of the high-frequency current demand locally, decreasing the time delay and the corresponding phase shift.

Circuit analysis quantifies the bypass benefit. The bypass capacitors reduce but do not eliminate the power-supply coupling effects. Voltage differences still develop with the residual impedance of the L_p, C_B combinations. The net line impedance Z_L reacts with signal current i_S, reducing the voltage magnitudes at both V_P and V_N by the amount $i_S Z_L$. Then, $V_P = V_+ - i_S Z_L$ and $V_N = V_- + i_S Z_L$, so the total supply voltage delivered to the op amp, $V_P - V_N = V_+ - V_- - 2i_S Z_L$, decreases by the amount $2i_S Z_L$. This reacts with the PSRR of the op amp, producing an amplifier input error of $e_{\text{idS}} = 2i_S Z_L/\text{PSRR}$. The circuit amplifies this error signal with the same A_{ne} gain that amplifies the op amp's input noise voltage to produce an output noise component

$$e_{\text{noS}} = \frac{2A_{\text{ne}} i_S Z_L}{\text{PSRR}}$$

Here, e_{noS} represents the output noise produced by supply-line coupling and for the generalized configuration of Fig. 8.1, $A_{\text{ne}} = 1 + R_2/R_1$.

8.1.2 The noise-coupling frequency response

Examination of the above expression reveals the photodiode amplifier's power-supply sensitivity and its frequency dependencies. The supply-coupled noise e_{noS} depends upon the line impedance Z_L, the amplifier's PSRR, and the noise gain A_{ne}. In the unbypassed state, each characteristic varies in a manner that increase e_{noS} as frequency increases. Comparison of unbypassed and bypassed cases reveals that supply bypass reduces the resulting e_{noS} response from a triple zero to the single zero normally expected with the photodiode amplifier. Unbypassed, a rising A_{ne} gain, a declining PSRR, and a rising Z_L each introduce zeroes in the e_{noS} response. Bypassed, a declining PSRR and a declining Z_L produce canceling effects and leave the single zero of the rising A_{ne}. Examination of these effects adds insight into potential noise origins for troubleshooting a circuit, especially in light of the practical bypass deficiencies described later.

As described before, the response zero of the A_{ne} noise gain results from the amplifier input error signal across the photodiode capacitance. This parallels the amplifier input noise voltage drop described in Sec. 5.1.2. In the figure, the $e_{idS} = 2i_S Z_L/\text{PSRR}$ error voltage drops across the photodiode and its capacitance. The resulting capacitive current $e_{idS}C_D s$ flows through feedback resistor R_f, producing the amplified output noise voltage, $e_{noS} = e_{idS}C_D sR_f = 2i_S Z_L C_D sR_f/\text{PSRR}$. As frequency increases, this error rises due to the numerator-based s term here. From a circuit perspective, this rise reflects the falling capacitive impedance and the increasing capacitive error current that results. The resulting output noise would simply follow the normal gain peaking effects of the Sec. 5.1.2 noise voltage analysis, but the other two response zeros increase the noise response further.

The second response zero results from the declining PSRR of the op amp. This decline increases the $e_{idS} = 2i_S Z_L/\text{PSRR}$ input error signal to in turn increase $e_{noS} = A_{ne}e_{idS}$. An approximation to the PSRR frequency response simplifies the mathematical expression of this effect. Over most of the amplifier's useful frequency range, PSRR typically rolls off with a single-pole response as shown in Fig. 8.2. In this single-pole span, the response drops with a 1:1 slope, making the PSRR decline in direct proportion to the increasing frequency. This declining curve crosses the unity gain, or 0-dB axis, at a crossover frequency f_{cp}, marking a reference point. Together, f_{cp} and the curve's 1:1 slope define PSRR $\approx f_{cp}/f = \omega_{cp}/s$ in this dominant single-pole range. Substituting this PSRR approximation in the preceding e_{noS} equation produces $e_{noS} = 2i_S Z_L C_D s^2 R_f/\omega_{cp}$. This expression displays the numerator-based s^2 term of a double-zero response, increasing high-frequency noise further.

Without bypass, a third response zero results from the inductances of the power-supply lines. Then, the line inductances make the line impedances $Z_L = L_p s$, and substituting this expression in the last e_{noS} equation produces $e_{noS} = 2i_S L_p C_D s^3 R_f/\omega_{cp}$. This expression displays the numerator-based s^3 factor of a triple-zero response. At higher frequencies, such a response produces a very large gain for the

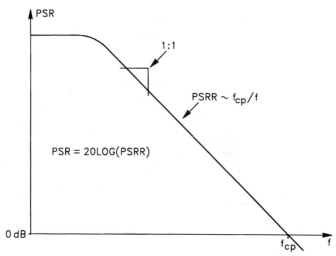

Figure 8.2 Op amp power-supply rejection displays a frequency response approximated by PSRR $\sim f_{cp}/f$, where f_{cp} is the unity-gain crossover of PSRR.

power-supply noise signal. Reducing this gain requires decreasing L_p and/or increasing ω_{cp}. Decreasing L_p reduces the supply-line noise signal, and increasing ω_{cp} improves the PSRR ability to reject this noise. These noise reduction options combine with power-supply bypass in the control of power-supply noise coupling.

Adding the C_B bypass capacitors of Fig. 8.1 dramatically reduces this noise coupling by removing two of the three zeroes in the e_{noS} response. Ideally, capacitive bypass dominates the higher-frequency line impedances, converting them from $Z_L = L_p s$ to $Z_L = 1/C_B s$. This removes a zero and adds a pole. Making this $Z_L = 1/C_B s$ substitution in the $e_{noS} = 2i_S Z_L C_D s^2 R_f/\omega_{cp}$ equation produces $e_{noS} = 2i_S C_D s R_f/C_B \omega_{cp}$. Here, a simple s factor in the numerator replaces the previous s^3, reflecting a single rather than triple zero for the e_{noS} response. Then, the power-supply noise produced by signal current i_S, and coupled to the amplifier input through PSRR, only receives the same amplification as the amplifier's input noise voltage. However, this bypass must also ensure frequency stability, and bypass deficiencies still require attention to the noise reduction options involving reduced L_p and increased ω_{cp} as described above.

8.1.3 Power-supply coupling and frequency stability

In addition to noise reduction, the power-supply bypass must serve a second and generally more prevailing purpose, the preservation of frequency stability. Lack of stability or oscillation produces the ultimate noise signal, independently sustaining itself and easily dominating the circuit's response error. As a result, the bypass selection generally focuses first upon stability and then upon improvements that further reduce noise coupling. Stability preservation requires bypass control of the parasitic feedback loop established by the supply-line impedance and the amplifier's PSRR coupling. Analysis of this feedback loop provides insight into the loop's origin and control and produces a mathematical expression of the condition required for oscillation.

As described with Fig. 8.1, this loop originates in the power-supply noise signal $2i_S Z_L$ that couples to the amplifier input in the error signal $e_{idS} = 2i_S Z_L/\text{PSRR}$. From there, the amplifier produces an output signal and a corresponding feedback error signal that actually supports e_{idS}. Figure 8.3 illustrates this feedback relationship with a modified version of the Fig. 8.1 circuit. As before, the circuit here neglects the input signal to focus the analysis upon signals related to the PSRR error. Modifications here alter the power-supply connection and the input error signal. For the supply connection, the amplifier load current i_L replaces the previous i_S, and the V_+ connection alone illustrates the supply coupling effect. Only the amplifier's i_L portion of i_S influences stability because only this portion produces a feedback relationship with the amplifier. Also, the coupling effects described here for the V_+ connection apply to those that occur with the V_- path as well. Next, a change in the input error signal shown reflects the effect of amplifier gain error. In the figure, the e_o/A gain error term appears in the more general input error

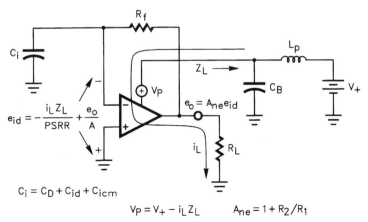

Figure 8.3 Load current i_L, drawn by the op amp, reacts with the supply-line inductance to produce an input error signal that in turn produces an output voltage and potential oscillation.

signal e_{id}, instead of the previous e_{idS}, to include the parasitic feedback mechanism.

8.1.4 The oscillation condition

Tracing a signal through the parasitic feedback loop illustrates both the feedback mechanism and its potential for oscillation. An intuitive evaluation illustrates the mechanism, and then a quantitative analysis defines the oscillation condition. In the figure, an e_o output voltage from the amplifier supplies a load current i_L to resistor R_L. The amplifier draws this current from V_+ and through the Z_L impedance of its supply line. The resulting line voltage drop produces a component of the e_{id} error signal, $-i_L Z_L/\text{PSRR}$, through the amplifier's finite PSRR. Then, the circuit amplifies this component by the amplifier noise gain A_{ne}, producing an e_o output response. That response reflects back to the amplifier inputs through the amplifier's open-loop gain A, producing the e_o/A component of the e_{id} shown. Thus, the power-supply coupling produces an input signal, which in turn produces an output signal that then produces an input signal. This simple link describes the full circle of a feedback loop capable of sustaining oscillation.

Quantifying this analysis produces two equations that describe the amplifier's feedback response and express the fundamental condition required for oscillation. First, the direct and secondary effects of power-supply coupling produce the input error signal $e_{\text{id}} = -i_L Z_L/\text{PSRR} + e_o/A$ as described above. Substituting $i_L = e_o/R_L$ removes the i_L variable for

$$e_{\text{id}} = \left(-\frac{Z_L}{R_L \text{PSRR}} + \frac{1}{A}\right) e_o$$

To sustain oscillation, e_{id} must independently support e_o through a second relationship. An op amp circuit amplifies its input error as expressed by[2]

$$e_o = \frac{A}{1+A\beta} e_{\text{id}} = A_{\text{ne}} e_{\text{id}}$$

where $A_{\text{ne}} = A/(1+A\beta)$ is the circuit's noise gain. The last two equations describe first the e_{id} dependence upon e_o and then a converse relationship. Oscillation results when the two signals support each other with no additional assistance. Combining the equations for the two produces the oscillation defining condition

$$e_o = e_o \frac{Z_L A - R_L \text{PSRR}}{(1+A\beta) R_L \text{PSRR}}$$

While too complex for intuitive evaluation, this last equation quickly yields two simplified results. With e_o on both sides of the equation,

Reducing Power-Supply Noise Coupling

only two solutions satisfy the equality expressed. First, $e_o = 0$ balances the equation, indicating the stable state with no self-sustaining oscillation signal. Second, dividing both sides of the equation by e_o and solving for Z_L defines the line impedance required to produce the oscillation state

$$Z_L = \frac{R_L \text{PSSR}}{A/(2+A\beta)} \approx \frac{R_L \text{PSRR}}{A_{ne}}$$

Preventing oscillation requires avoiding conditions that balance this new equality. Keeping Z_L low avoids oscillation as long as

$$Z_L < \frac{R_L \text{PSRR}}{A_{ne}}$$

This last Z_L equation offers several insights into the bypass requirement and the related parameters that control stability. Variations in three parameters of the equation influence the Z_L requirement. At lower frequencies, bypass capacitors fail to reduce Z_L, increasing the value of the left side of the equation. However, at these low frequencies, PSRR remains high, increasing the right side and preserving the equation's $<$ condition. At higher frequencies, PSRR declines, decreasing the value of the right side. However, bypass capacitors then reduce Z_L to retain the stability condition. At all frequencies, the value of load resistance R_L influences stability. As expressed above, lower values of R_L require correspondingly lower values of Z_L. There, lower R_L values produce greater supply-current drains that increase the supply-voltage drop developed on the Z_L impedance.

The above analysis provides intuitive but simplified guidelines to stability preservation through control of the line impedance's magnitude. In practice, phase shifts also influence stability, and complex phase conditions actually accompany the Z_L magnitude requirement above. The PSRR roll off adds a 90° phase shift, and the bypassed Z_L impedance displays multiple resonances that potentially produce 180° phase transitions. These transitions compromise stability in a manner not evident from cursory examination of the above Z_L equation. The complexity of the overall result presents a formidable analysis task again best avoided by design practices intuitively communicated by that Z_L equation. Then, examination of the Z_L resonances produces simple design equations that ensure stability.

8.2 Selecting the Primary Bypass Capacitor

As described above, supply bypass that maintains low line impedance limits supply noise coupling and preserves frequency stability. This suggests making the bypass capacitance large to minimize the resulting $Z_L = 1/C_B s$ impedance. However, power-supply bypass does not simply replace the $Z_L = L_p s$ impedance with $Z_L = 1/C_B s$ as previously approximated. This approximation fails due to a variety of resonances formed with parasitic inductances. The basic power-supply bypass produces two such L-C resonances. Just the line inductance and the basic bypass capacitor produce the first, and the capacitor itself produces the second. The first one must be placed within the useful frequency range of the op amp in order to realize the bypass benefit. However, the second one can often be placed outside this range by avoiding an excessively large capacitance value. Larger capacitors have greater internal inductances that move this second resonance to lower frequencies. A compromise solution produces a design equation that adequately reduces line impedance but keeps this second resonance at a higher frequency. Later, the addition of a secondary bypass capacitor contends with the second resonance for higher-frequency amplifiers that cannot escape it.

8.2.1 Bypass resonances

The first results from the reaction of C_B with L_p and occurs well within the frequency range of the amplifier. In Fig. 8.3, C_B effectively appears in parallel with the L_p supply-line inductance. There, the ideal zero impedance of V_+ returns L_p to ac ground, completing this parallel connection. Then, C_B and L_p form the classic L-C tank circuit with an impedance of

$$Z_L = \frac{L_p s}{1 + L_p C_B s^2}$$

At low frequencies, $Z_L = L_p s$, and at high frequencies, $Z_L = 1/C_B s$ as previously approximated. However, at an intermediate frequency, this impedance displays a resonance maximum at the frequency $f_{rL} = 1/2\pi \sqrt{L_p C_B}$. There, the Z_L line impedance approaches infinity, totally counter to the impedance reduction intended by the power-supply bypass.

Fortunately, a parasitic line resistance R_p and the amplifier's PSRR contend with this resonance. The R_p resistance dissipates the resonant energy of the L-C tank, detuning it to dramatically restrain the resulting impedance rise. A typical printed circuit trace introduces a parasitic 12 mΩ per inch in its connection between the power supply and the amplifier. In addition, choosing C_B to place this resonance at a lower frequency where PSRR remains high lets the amplifier attenuate the associated coupling effect. Figure 8.4 illustrates the frequency response of this Z_L impedance along with the impedance's

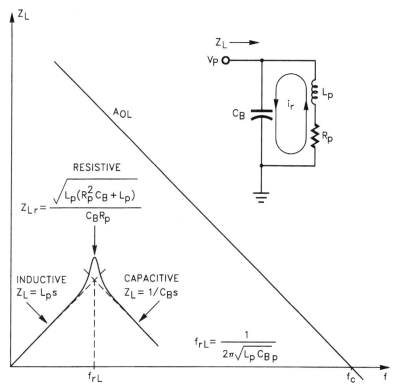

Figure 8.4 Together, the power-supply line inductance and a bypass capacitance form an L-C tank circuit with a resonant impedance detuned only by the presence of parasitic line resistance R_p.

circuit model. Superimposed on the figure, the amplifier's A_{OL} response provides a relative frequency comparison. In the model, bypass capacitor C_B shunts the series combination of the parasitic line inductance L_p and the parasitic line resistance R_p. This model assumes the zero output impedance of an ideal power supply and returns the L_p, R_p combination to ground. Then, the circuit model represents the Z_L line impedance seen from the V_P terminal of the amplifier's positive supply connection.

As shown by the Z_L response plot, the line impedance varies from inductive to resistive to capacitive as frequency increases. At lower frequencies, line inductance L_p controls Z_L, producing a rising impedance curve. At higher frequencies, bypass capacitor C_B takes control, producing the declining Z_L curve desired. Between these two frequency ranges, the inductive and capacitive response curves cross at their resonant frequency f_r. There, Z_L rises to a peak value and actually increases because of the presence of bypass capacitor C_B. Fortunately, the presence of resistance R_p limits this rise by detuning the tank circuit.

8.2.2 An intuitive evaluation of bypass resonance

Multiple supply-line resonances potentially affect an op amp circuit. An intuitive evaluation of the underlying L-C resonance adds insight into their cause and control. For this evaluation, consider the component curves that define the Fig. 8.4 Z_L response. As mentioned, inductive and capacitive influences largely shape this response with one or the other dominant at frequencies separated from the resonance at f_r. However, in the vicinity of f_r, the magnitude and phase relationships of the inductive and capacitive components combine to increase the Z_L line impedance. The effect of this combination parallels the magnitude/phase relationship of the basic stability issue described in Chap. 3. There and here, the crossing of two magnitude response curves marks the frequency of magnitude equality since the two curves then occupy the same point on a graph. In the figure, dashed line extensions of the inductive and capacitive impedance responses mark this point where the $Z_L = L_p s$ and $Z_L = 1/C_B s$ lines cross at f_r.

At this crossing, the slopes of the two lines reflect their associated phase shifts and the slope difference, or rate of closure, reflects the corresponding phase difference. At f_r, this phase difference turns this equal magnitude point into a resonance. The single-zero rising response of the $L_p s$ component reflects a 90° phase lead, and the single-pole declining response of the $1/C_B s$ component reflects a 90° phase lag. This combination produces a 180° phase difference between the impedances of the circuit model's L_p and C_B components. For the ideal L-C tank circuit, this magnitude and phase combination produces an infinite impedance, and any supply of energy to the tank produces oscillation. Simply turning on the power supply provides this energy through the transient current drawn from the supply. After that, the ideal tank's infinite impedance successfully sustains an oscillation signal even after termination of energy input to the circuit.

Initially neglecting R_p of the Fig. 8.4 circuit model illustrates the resonant condition. Then, the parallel connected L_p and C_B support the same voltage drop. At the resonance, the equal impedance magnitudes of L_p and C_B develop currents of equal magnitudes from their common voltage drop. Yet, the 180° phase difference for the two impedances makes the two currents out of phase or of opposite polarity. This produces the circulating resonant current i_r shown with no need for additional energy input, i.e., oscillation. Current i_r flows through C_B in one direction and through L_p in the other, yet the two flows develop the same voltage drop, as required by the parallel connection of these elements. Thus, a self-sustaining i_r loop results in which energy transfers from the inductor to the capacitor and back, developing a sinusoidal voltage signal on Z_L at the resonant frequency f_r.

Fortunately, the presence of parasitic resistance R_p introduces energy dissipation to reduce the resonant impedance and stop the oscillation. Analysis of the figure's simple R-L-C combination shows that the line impedance peaks at the resonance frequency f_r. There, Z_L displays the 0° phase shift of resistive impedance and a resonant magnitude of

$$Z_{\text{Lr}} = \frac{\sqrt{L_p \left(R_p^2 C_B + L_p\right)}}{C_B R_p}$$

Examination of this equation confirms anticipated effects. Z_{Lr} requires nonzero values for both C_B and R_p to prevent an infinite peak impedance. Also, making C_B large decreases this impedance and reduces its resonant impedance equation to $Z_{\text{Lr}} = \sqrt{L_p/C_B}$.

8.2.3 The bypass selection

Control of the above resonance and an empirical guideline produce the first two design equations for the bypass selection. The first equation places the resonance at a frequency where a higher-amplifier PSRR better contends with the associated impedance rise. The second equation sets the magnitude of the bypassed line impedance with a C_B design equation. Later, control of other resonances completes the process with secondary bypass capacitor selection.

At first, the impedance increase of the Z_{Lr} resonant condition would suggest moving f_{rL} to a frequency above the normal operating frequency range of the op amp, f_c. Here, f_c represents the unity-gain crossover frequency of the op amp in Fig. 8.4 and generally represents the upper limit of the amplifier's useful frequency range. Above f_c, the lack of amplifier gain restricts the parasitic feedback resulting from the resonance. However, making $f_{\text{rL}} > f_c$ would entirely sacrifice the benefit of the supply bypass. Preceding f_{rL} in Fig. 8.4, the Z_L impedance rises with frequency, unaffected by the bypass. The Z_L impedance only reflects the desired decline after the f_{rL} resonance where the capacitive roll off dominates Z_L. Thus, the f_{rL} resonance must be endured within the op amp's operating frequency range in order to benefit from the higher-frequency effects of the bypass.

Moving this bypass resonance to a frequency $f_{rL} \ll f_c$ actually provides the best compromise. This places the resonance at a frequency where the amplifier retains sufficient PSRR to attenuate the coupling effect of the resonance. Further, the $f_{rL} \ll f_c$ placement moves the C_B roll off of the line impedance back into the amplifier's useful frequency range. Then, for $f_{rL} = 1/2\pi\sqrt{L_p C_B} \ll f_c$ the C_B capacitance value must satisfy the limit equation

$$C_B \gg \frac{1}{4L_p (\pi f_c)^2}$$

For a typical case, $L_p \approx 300$ nH and $f_c \approx 2$ MHz, producing $CB \gg 0.02$ μF. Generally, the line impedance control described next automatically satisfies this C_B condition.

Making $f_{rL} \ll f_c$ also reduces the analysis of other bypass conditions to the impedance of the capacitor alone. Without bypass, the line impedance actually includes a complex combination of inductances and resistances introduced by wire, printed circuit traces, and connectors. All of these affect the lower-frequency response of the Z_L impedance, discouraging any detailed analysis. However, making $f_{rL} \ll f_c$ reduces the Z_L impedance to just that of the capacitor in the critical higher-frequency range. There, a declining PSRR most requires a low Z_L to limit supply-line coupling. Above the f_{rL} resonance Z_L then becomes $Z_L = Z_{CB} = 1/C_B s$ independent of the preceding line impedance complexity.

Then, for most op amps, simply reducing the C_B impedance to about 1 Ω well before f_c effectively counteracts the line impedance effects. This reduces the previous stability condition to $Z_L = Z_{CB} = 1 < R_L \text{PSRR}/A_{ne}$ or $A_{ne} < R_L \text{PSRR}$. Here, a noise gain as large as 1000 and a load resistance as small as 1 kΩ permit a PSRR as low as unity before oscillation results. Rarely would such a low PSRR occur where the noise gain remained this high, so the $Z_{CB} = 1$ Ω guideline assures stability for virtually every case. Setting $Z_{CB} = 1/2\pi f\, C_B = 1$ Ω produces $C_B = 1/2\pi f$. Then, setting $f = f_c/100$ for this 1-Ω condition produces a general design equation for C_B

$$C_B = \frac{50}{\pi f_c}$$

For a general-purpose op amp with $f_c \approx 2$ MHz, this equation defines a capacitance value of $C_B \approx 7.5$ μF. Retaining this 1-Ω impedance up to and beyond f_c requires additional bypass attention as described below.

8.3 Selecting a Secondary Bypass Capacitor

Capacitor self-resonance potentially compromises the bypass described above, requiring the addition of secondary bypass capacitors. For general-purpose op amps, the above bypass selection generally places this self-resonance at a frequency beyond the amplifier's response range. Higher-frequency amplifiers encompass this frequency and may require the additional bypass. Still, even those capacitors exhibit self-resonances and require careful selection through a compromise that again produces simple design equations.

8.3.1 Bypass capacitor self-resonance

The inherent parasitic inductances and resistances of capacitors also affect the bypass effectiveness. The inductances introduce new resonances, and the resistances limit the line impedance reduction. A first new resonance results just from the capacitance and inductance of the primary bypass capacitor in a self-resonance condition. At this resonance, the bypass impedance would drop to zero except for the presence of the parasitic resistance. Above that frequency, the capacitor's parasitic inductance overrides the intended capacitance in the control of the bypass impedance.

Making $f_{rL} \ll f_c$ in Sec. 8.2 above typically requires a fairly large capacitance value in the multiple microfarad range. Capacitors this large introduce a new resonance condition that compromises the bypass effectiveness at frequencies typically within the amplifier's response range. Such cases sometimes require the addition of a secondary, smaller bypass capacitor as described later. The new resonance results from the parasitic inductance L_{Bp} of the bypass capacitor reacting with the intended capacitance. This resonance reverses the Z_{CB} response slope as shown in Fig. 8.5. All capacitors possess this internal inductance, with a value corresponding largely to the lengths of the capacitor's external and internal connecting paths. Minimizing the total connecting length minimizes the parasitic inductance and maximizes the frequency of the self-resonance. First, just minimizing the capacitor lead length, including that of associated circuit board traces, addresses the external path length. Then, capacitor selection minimizes the internal path component as primarily determined by the capacitor's construction material. For a given capacitance value, capacitor materials with higher capacitance densities reduce this length. Tantalum, NPO ceramic, and mica capacitors offer the lower inductance alternatives for high, medium, and low capacitance values, respectively.

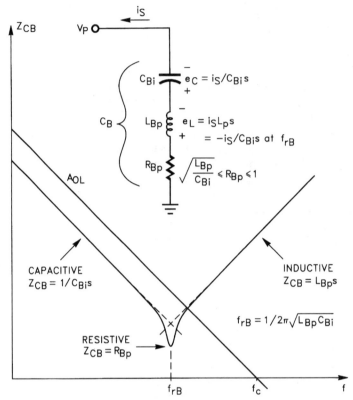

Figure 8.5 The intended impedance decline of a bypass capacitor reverses when the capacitor's internal inductance takes control at higher frequencies.

More detailed examination of the bypass capacitor's actual impedance components provides insight into its nonideal behavior and into the means to control the bypass result. As modeled in Fig. 8.5, the capacitor's parasitic inductance appears in series with the intended capacitance along with a parasitic resistance. There, L_{Bp} produces the self-resonance with C_B's ideal capacitance, C_{Bi}. Parasitic resistance R_{Bp} sets the capacitor's impedance at the resonance. This resistance results from the same connecting path that produces the capacitor's inductance. Supply-current drain i_S develops voltages across each of the model's impedances with the capacitive and inductive impedances producing canceling effects at their resonance. Current i_S produces the voltage drops $e_C = i_S/C_{Bi}s$ across C_{Bi}, and $e_L = i_S L_{Bp} s$ across L_{Bp}. At the resonance, $s = s_{rB} = j\omega_{rB} = j/\sqrt{L_{Bp}C_{Bi}}$. Substituting this result for s in the preceding voltage drop equations yields $e_{Cr} = -j\sqrt{L_{Bp}/C_{Bi}}$ and $e_{Lr} = j\sqrt{L_{Bp}/C_{Bi}}$. Thus, at the resonance, $e_{Cr} = -e_{Lr}$ and the two voltage drops cancel. Except for the presence of R_{Bp}, this voltage cancellation would produce zero voltage drop in response to the i_S current at the resonant frequency, indicating the ideal zero bypass impedance at the resonance.

However, parasitic resistance R_{Bp} produces a third voltage drop with i_S and limits the impedance decline to the level of this resistance. At first, this limit would seem undesirable, but the parasitic resistance also detunes the C_{Bi}, L_{Bp} resonance, taming a dramatic phase transition in the line impedance. That transition presents a broad range of phase conditions that could degrade stability. In general, this effect rarely produces a significant stability disturbance because it occurs at a frequency where amplifier PSRR remains relatively high. However, the impedance rise following the resonance may well degrade stability and require dual bypass. For the general case, the beneficial effect of the capacitor's parasitic resistance provides adequate stability assurance and eases the capacitor selection. Accepting a capacitor with higher parasitic resistance actually smoothes the phase transition of the resonance.

This resistance benefits performance as long as it detunes the resonance but does not raise the line impedance above the 1-Ω guideline level. For minimum detuning, the impedances of the R, L, and C components of the tank circuit should be equal at the resonant frequency. At this frequency, the impedance magnitudes of C_{Bi} and L_{Bp} automatically equate as shown in Fig. 8.5. There, dashed-line extensions of the two impedance curves identify equal magnitudes at their f_{rB} crossing. At this crossing, $Z_{CBi} = 1/2\pi f_{rB} C_{Bi}$, where $f_{rB} = 1/2\pi \sqrt{L_{Bp} C_{Bi}}$. Setting R_{Bp} equal to the Z_{CBi} impedance and solving produces the minimum resistance required for detuning, $R_{Bp} = \sqrt{L_{Bp}/C_{Bi}}$. Combining this minimum with the 1-Ω maximum produces the design limit equation

$$\sqrt{\frac{L_{Bp}}{C_{Bi}}} \leq R_{Bp} \leq 1$$

For the typical case, $L_{Bp} = 15$ nH and $C_{Bi} = 7.5$ μF, making the limit condition an easily met $0.04 \leq R_{Bp} \leq 1$. Later, dual bypass conditions raise this resistance limit to counter yet another resonance.

Op amps having lesser signal bandwidths generally assure this detuned condition in the bypass capacitor selection. Such amplifiers require larger C_B values for the previous C_B design equation, directing the capacitor selection to tantalum types. Such capacitors display larger R_{Bp} values that flatten the Fig. 8.5 transition between its capacitive and inductive slopes. However, for higher-frequency amplifiers this guideline would permit use of ceramic capacitors which display much lower R_{Bp} values. Then, replacing an initial ceramic choice with tantalum could solve occasional, unexpected stability problems. Later, a resistive detuning evaluation explores this option further.

Following the f_{rB} resonance, the Z_{CB} impedance curve rises in Fig. 8.5, again threatening stability. However, the C_B selection guidelines described above typically control the effect of this rise for general-purpose op amps. There, the avoidance of excessively large capacitance values and the limited amplifier bandwidths reduce circuit susceptibility to the impedance increase. Choosing $C_B = 50/\pi f_c$ assures adequate line impedance control at lower frequencies without introducing the excess inductance of a larger capacitor. With this compromise the Z_{CB} impedance rise remains within the 1-Ω limit over the useful bandwidth of general-purpose amplifiers. Success here also depends upon minimizing the connection length between the capacitor and the amplifier.

8.3.2 Dual bypass capacitors

The simple, single-capacitor bypass that serves general-purpose op amps fails with higher-frequency op amps. At higher frequencies, inductive effects become far more serious, requiring greater attention to frequency stability and supply noise coupling. The parasitic inductance of the larger-value, single bypass capacitor defeats the bypass purpose at higher frequencies, requiring the addition of a smaller secondary capacitor. Even general-purpose amplifiers may require dual bypass if PSRR is low.

In Fig. 8.5, the parasitic L_{Bp} inductance controls the bypass impedance Z_{CB} at frequencies above the capacitor's self-resonance. At higher frequencies, the rising $L_{Bp}s$ portion of Z_{CB} increases the bypass impedance, reducing the bypass effectiveness. This rise generally begins within the amplifier response range and, there, a declining PSRR makes the amplifier more vulnerable to the line impedance. For all but the highest-frequency applications, just adding a secondary bypass capacitor in parallel with the first restores a low bypass impedance for the full amplifier response range. This produces a net $C_B = C_{B1} + C_{B2}$, and making $C_{B2} \ll C_{B1}$ generally assures that the C_{B2} self-resonance occurs outside the amplifier's response range. Both the lower capacitance value of C_{B2} and the accompanying lower parasitic inductance produce this higher resonant frequency. For even higher-frequency applications, yet a third capacitor may be required. In either case, examination of the dual bypass case here illustrates the impedance characteristics that guide the bypass selection process.

Adding secondary bypass capacitors restores the declining frequency response of the Z_{CB} bypass impedance but with new complications.

This solution introduces two additional resonances: one from the self-resonance of the secondary capacitor and the other from the reaction of the second capacitance with the inductance of the first. Figure 8.6 illustrates the first new resonance with a circuit model and the corresponding impedance responses. In the model, both capacitors display parasitic inductances and resistances and all of these parasitics affect the response curves. In the graph, two curves of the figure display the resulting capacitor impedances as Z_{CB1} and Z_{CB2}. At lower frequencies, a declining Z_{CB1} provides the lower impedance bypass shunt. However, Z_{CB1} later resonates and begins to rise at f_{rB1}. At higher frequencies, Z_{CB2} bypasses this rise and initially restores a declining bypass impedance. Later, the self-resonance of C_{B2} at f_{rB2} again produces a rise but at a lower impedance level than provided by the Z_{CB1} curve. While the resonances remain inevitable, their careful placement optimizes the overall bypass effect. To aid this placement, a third curve of the graph shows the amplifier's open-loop gain response A_{OL}. Comparison of the graph's three responses guides the secondary capacitor selection in a manner that ensures the amplifier's frequency stability.

8.3.3 Dual bypass selection

Selection of the primary C_{B1} capacitor indirectly defines the C_{B2} capacitance value through impedance interaction. The fundamental bypass requirement defines C_{B1} as described before for the single bypass case: $C_{B1} = 50/\pi f_c$. This selection reduces the line impedance to 1 Ω at a frequency well within the amplifier's response range. Then, the addition of C_{B2} bypasses the higher-frequency impedance rise produced by the L_{Bp1} inductance of C_{B1}. Selecting C_{B2} to limit this rise to the same 1-Ω guideline level defines this secondary capacitor's value.

In Fig. 8.6, the net bypass impedance of the C_{B1}, C_{B2} parallel combination peaks at the f_{iB} intercept of the rising Z_{CB1} curve and the falling Z_{CB2}. For higher-frequency amplifiers this peak would typically occur within the amplifier's response range. Selecting a value for C_{B2} that limits this peak to 1-Ω continues the stability preservation described before for the C_{B1} selection. At their f_{iB} intercept, the two Z_{CB} curves occupy the same point on the graph, making $Z_{CB2} = Z_{CB1}$. There, inductance controls $Z_{CB1} = 2\pi f_{iB} L_{Bp1}$ and capacitance controls $Z_{CB2} = 1/2\pi f_{iB} C_{Bi2}$. Equating the two impedance expressions defines the intercept frequency as

$$f_{iB} = 1/2\pi \sqrt{L_{Bp1} C_{Bi2}}$$

Thus, the parasitic inductance of C_{B1} and the ideal capacitance of C_{B2} combine to define the frequency of the bypass impedance peak.

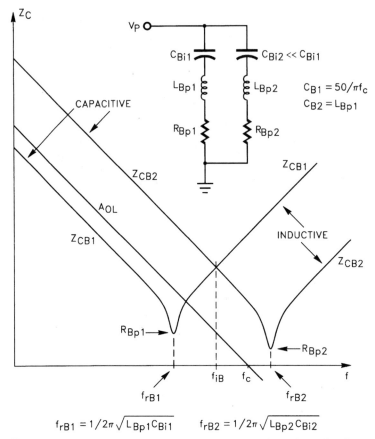

Figure 8.6 Adding a second bypass capacitor of smaller value rolls off the inductive impedance rise of the first.

Translating this f_{iB} expression into a C_{B2} design equation requires one further step. At f_{iB}, the Z_{CB2} curve follows its capacitive roll off as expressed by $Z_{CB2} = 1/2\pi f\, C_{B2}$. Setting $Z_{CB2} = 1\ \Omega$ and $f = f_{iB}$ in this expression produces $f_{iB} = 1/2\pi\, C_{B2}$. Equating this result with the above f_{iB} expression yields the C_{B2} design equation

$$C_{B2} = L_{Bp1}$$

Thus, simply making the magnitude of the C_{B2} capacitance equal to that of C_{B1}'s parasitic inductance transfers line impedance control from Z_{CB1} to Z_{CB2} at the 1-Ω level. In practice a resonance described later raises the peak impedance above 1 Ω at f_{iB}, but temporarily ignoring this effect yields the best starting point for the bypass selection. As described later, resistive detuning of this resonance or an alternate capacitor type solves any related problem without altering the above design equation. Using this equation, a typical 1.5 nH of C_{B1} inductance prescribes a secondary bypass of $C_{B2} = 1.5$ nF.

The above C_{B2} equation requires knowing L_{Bp1}, and impedance measurement provides the most reliable determination of this inductance. Capacitor data sheets rarely specify this parasitic. Some data sheets communicate this inductance through an impedance-versus-frequency curve. There, the higher-frequency, rising portion of this curve yields the parasitic inductance through the simple calculation $L_p = Z_C/2\pi f$. Unfortunately, even these curves can fail to communicate the actual inductance condition of practical applications. For packaged capacitors, this resonance includes the parasitic inductance of package leads that are largely removed upon installation. Chip capacitors avoid this lead effect, but their response curves still do not include the effect of the circuit board traces that connect the capacitor. Fortunately, today's impedance analyzers accurately measure the frequency response of a given capacitor and its approximated connection length. From such measurements, the rising portion of the impedance curve then accurately defines the actual inductance through the $L_p = Z_C/2\pi f$ relationship.

8.4 Bypass Alternatives

The preceding sections outline the bypass selection that adequately serves most applications. However, three special conditions may require additional attention to supply-line coupling. For these conditions, a third bypass capacitor, filtering, or resistive detuning further reduce the coupling effects. The third capacitor alternative extends the frequency range of bypass control for even higher-frequency amplifiers. There, the amplifier's A_{OL} response can encompass the self-resonances of both C_{B1} and C_{B2}. Then, adding a third, and even smaller, bypass capacitor rolls off the inductive impedance of C_{B2}, just as C_{B2} did for C_{B1}. Amplifiers with bandwidths exceeding 30 MHz may require this third capacitor.

A second alternative, filtering, removes supply-coupled signals outside the useful PSRR range of the amplifier. There, coresident circuitry may introduce high-frequency supply-line signals that couple through the amplifier with little PSRR attenuation. Beyond the PSRR's unity-gain crossover, such signals typically couple straight through the amplifier to its output. However, this PSRR crossover typically resides near the amplifier's A_{OL} crossover, presenting a filtering opportunity. The A_{OL} crossover marks the endpoint of the amplifier's useful frequency range. Thus, low-pass filtering, applied after the amplifier, can often remove the effect of high-frequency PSRR coupling without restricting the useful bandwidth of the application. Still, intermodulation effects can circumvent the filtering as described later.

Resistive detuning of the bypass impedance offers a third alternative for the reduction of supply coupling effects. The dual bypass configuration potentially produces a critical resonance that degrades stability at surprising frequencies. This may occur at a frequency below or above the amplifier's response crossover at f_c. Below f_c, the C_{B2} selected above restores a line impedance below the 1-Ω guideline level, presumably preserving stable operation. However, the C_{B2} resonance with L_{Bp1} can raise the net line impedance well above this level, producing oscillation. In other cases, the resonance can even produce oscillation at frequencies above f_c. There, diminished amplifier gain limits the parasitic feedback loop and would seem to prevent instability. However, the resonant impedance rise can counteract this limit. When necessary, resistive degeneration detunes this new resonance, and an appropriate capacitor selection generally provides this through parasitic resistance.

8.4.1 Detuning the dual bypass resonance

An interaction between the two capacitors of the dual bypass configuration develops this new resonance. There, the secondary capacitance reacts with the inductance of the primary in another L-C tank configuration. Figure 8.7 illustrates this effect with the bold Z_{CB} response curve representing the net impedance of the parallel-connected bypass capacitors. There, Z_{CB} first follows the Z_{CB1} curve of C_{B1} at lower frequencies and later follows the Z_{CB2} curve of C_{B2}. The transition between the two occurs at their f_{iB} intercept where the bypass combination produces a resonance of its own. This resonance increases the Z_{CB} magnitude as shown and produces a 180° transition in the phase of Z_{CB}. The magnitudes and phases of the amplifier's PSRR and A_{OL} responses combine with these Z_{CB} characteristics to produce a complex parasitic feedback condition.

Some combinations of application conditions make the magnitude and/or phase of this resonance degrade stability. In such cases, attention to the more conventional stability determinant, the amplifier's $1/\beta$ intercept, fails to identify the problem. Similarly, adding a third bypass capacitor produces no improvement. The third capacitor only reduces the line impedance at frequencies above this interaction resonance. For such cases, bypass degeneration moderates the PSRR error signal by reducing both the magnitude and phase shift of the supply-line impedance. Careful selection of the degeneration resistance improves stability without increasing the supply-line impedance in the critical higher-frequency ranges.

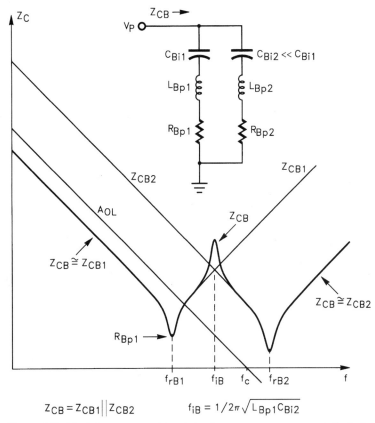

Figure 8.7 Dual bypass capacitors introduce a new resonance at their f_{iB} intercept where C_{Bi2} reacts with L_{Bp1}, increasing the net Z_{CB} impedance.

$$Z_{CB} = Z_{CB1} \| Z_{CB2} \qquad f_{iB} = 1/2\pi\sqrt{L_{Bp1}C_{Bi2}}$$

Graphical evaluation of this f_{iB} resonance reveals its underlying causes and displays the cure provided by the resistive degeneration. The resonance results from the reaction of C_{B1}'s parasitic inductance with C_{B2}'s capacitance. At f_{iB} in Fig. 8.7, Z_{CB1} follows a rising impedance response as controlled by L_{Bp1}, and Z_{CB2} follows a declining response as controlled by C_{Bi2}. At the crossing of the two responses, resonance results due to the equal magnitudes and opposite phases of the Z_{CB1} and Z_{CB2} impedances. This crossing or intercept of the two curves marks the point of equal impedance magnitudes since both curves occupy the same point in the graph. There, neither impedance dominates the parallel combination, in contrast to the impedance conditions at points separated from f_{iB}. At this intercept, the equal magnitudes of Z_{CB1} and Z_{CB2} contribute equally to the combined Z_{CB} impedance, making the phase shifts of the two also equally significant.

The single-zero rise of the Z_{CB1} curve corresponds to a 90° phase lead, and the single-pole decline of Z_{CB2} corresponds to a 90° phase lag. The 180° phase difference and equal impedance magnitudes inherently produces a resonance like that described earlier in Sec. 8.2.2.

Adding a small resistance in series with C_{B1} detunes this resonance to ensure stability. In practice, just choosing a different capacitor type for C_{B1} often serves this purpose through the capacitor's parasitic resistance. This added resistance does increase the Z_{CB} magnitude in a lower-frequency range, but it then decreases Z_{CB} in the higher, more critical frequency range of f_{iB}. Figure 8.8 illustrates the bypass optimum with degeneration resistance R_{D1} added in series with the other impedance elements of C_{B1}. As in Fig. 8.7, the bold Z_{CB} curve begins by following Z_{CB1} and ends by following Z_{CB2}. Unlike Fig. 8.7, the Z_{CB} curve now makes a gradual rather than resonant transition between Z_{CB1} and Z_{CB2} at their f_{iB} intercept.

The addition of R_{D1} actually detunes two resonances, first the self-resonance of C_{B1} and then the interactive resonance of C_{B1} and C_{B2}. Comparison of Fig. 8.8 with Fig. 8.7 reveals the overall effect of this resistance. Previously, the self-resonance of C_{B1}, at f_{rB1} in Fig. 8.7, produced a resonant impedance drop to the level of parasitic resistance R_{Bp1}. In Fig. 8.8, the addition of R_{D1} removes this drop and raises this level to $R_{D1} + R_{Bp1}$. This actually raises the bypass impedance in the region of the previous f_{rB1} resonance. However, the amplifier generally retains a high PSRR in this frequency range for sufficient attenuation of supply-line effects. Further, the Z_{CB} curve now makes a smooth rather than resonant transition to this new limit level. There, the reduced Z_{CB} response slope indicates a greatly reduced phase transition at the frequency of the previous f_{rB1} resonance. This reduces the potential phase combinations with amplifier gain and PSRR effects that could degrade stability. Accepting this temporary Z_{CB} increase allows R_{D1} to detune the more critical resonance at f_{iB}. There, the Z_{CB} curve no longer displays a resonant peak, indicating a greatly reduced Z_{CB} magnitude and phase transition at f_{iB}.

8.4.2 Selecting the detuning resistance

Detuning resistance R_{D1} should be large enough to prevent the resonant impedance rise at f_{iB} but not so large as to unnecessarily raise the Z_{CB} impedance at lower frequencies. Graphical analysis of Fig. 8.8 reveals this optimum condition and defines an R_{D1} design equation. The curves of the figure identify this optimum through the magnitudes and slopes of the impedance responses.

Detuning the Fig. 8.7 resonance at f_{iB} requires reducing the slope difference, or rate of closure, of the Z_{CB1} and Z_{CB2} curves at their f_{iB} in-

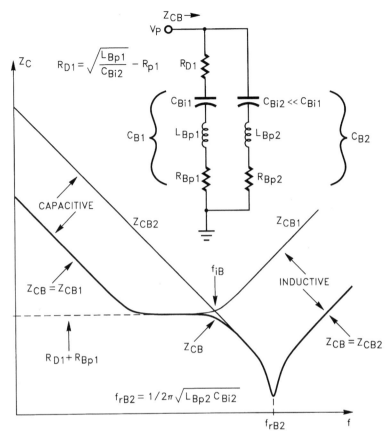

Figure 8.8 Adding resistance R_{D1} in series with C_{B1} detunes two of the Fig. 8.7 resonances at the expense of an elevated Z_{CB} level at lower frequencies.

tercept. This rate-of-closure criteria follows from the Chap. 3 stability analysis where the slopes of intersecting response curves reflect phase differences and the potential for circuit oscillation. Here, the slopes of the Z_{CB1} and Z_{CB2} curves similarly reflect phase differences and the potential for resonance. Adding R_{D1} decreases this slope difference at the f_{iB} intercept of Fig. 8.8, detuning the resonance. This added resistor produces the level zero-slope portion of the Z_{CB1} response, and this slope reflects zero phase shift. At the f_{iB} intercept, Z_{CB2} still follows its single-pole roll off, reflecting a 90° phase lag for a net 90° phase difference at the intercept. This greatly reduces the phase difference from the 180° required for resonance and would absolutely remove the resonant reaction of Z_{CB1} and Z_{CB2}.

However, bypass optimization produces a compromise favoring a somewhat greater phase difference at f_{iB}. The $R_{D1} + R_{Bp1}$ resistance, which produces the level region of the Z_{CB1} response, eventually looses control, returning the response curve to a single-zero rise. In the Fig. 8.8 example, the L_{Bp1} impedance overrides that of R_{D1} just at the f_{iB} intercept. Phase shift accompanies the following rise, and intuition first suggests increasing the R_{D1} resistance to move the Z_{CB1} rise, and its phase shift, well beyond the f_{iB} intercept. However, such a choice would unnecessarily increase the impedance magnitude of Z_{CB}'s new level region at $R_{D1} + R_{Bp1}$. There, a continued increase of R_{D1} would eventually increase rather than decrease the impedance at f_{iB}. In compromise, setting the $R_{D1} + R_{Bp1}$ level at the intercept magnitude still detunes the resonance while restraining the impedance increase of the level region.

As illustrated, this optimum level places the zero of the Z_{CB1} rise at the f_{iB} intercept. Then, Z_{CB1} introduces 45° rather than the ideal 0° of phase shift at the intercept. This increases the net phase difference with Z_{CB2} from 90° to 135° but still avoids the 180° that produced the resonance before. For this compromise, $R_{D1} + R_{Bp1} = Z_{CB2}$ at f_{iB}. Then, $R_{D1} + R_{Bp1} = 1/\omega_{iB} C_2$, where $\omega_{iB} = 2\pi f_{iB} = 1/\sqrt{L_{Bp1} C_{B_2}}$. Combining these expressions and solving for R_{D1} yields $R_{D1} = \sqrt{L_{Bp1}/C_{B2}} - R_{Bp1}$. From before, the 1-Ω impedance guideline defines $C_{B2} = L_{Bp1}$, and substitution yields the design equation

$$R_{D1} = 1 - R_{Bp1}$$

Expressed differently, this equation prescribes $R_{D1} + R_{Bp1} = 1$ Ω for a total 1-Ω resistance in the C_{B1} path. This repeats the 1-Ω guideline underlying bypass selection.

Typically, the above R_{D1} equation prescribes difficult resistance values of a fraction of an ohm. Such resistors exist but can be expensive and more difficult to acquire. However, when required, only this solution removes the effects of the f_{iB} bypass resonance. Fortunately, just choosing a different capacitor for C_{B1} generally provides an adequate detuning resistance. Choosing a C_{B1}' with $R_{Bp1}' \approx 1$ Ω detunes the f_{iB} resonance but still avoids exceeding the 1-Ω guideline. Fortunately, this resistance does not have to be precise to provide sufficient detuning of the f_{iB} resonance. Often, just switching C_{B1} from a ceramic type to tantalum serves this purpose.

Note that this C_{B1} switch also requires changing the secondary bypass value C_{B2}. A tantalum C_{B1}' substitutes an increased L_{Bp1}' for L_{Bp1}, requiring recomputation of the corresponding $C_{B2}' = L_{Bp1}'$. Increasing C_{B2} here theoretically increases its inductance and would degrade higher-frequency stability. However, practical limitations generally leave the net inductance condition unchanged. In practice, some minimum lead length and inductance always accompany a capacitor construction and installation. Just the lead length from an amplifier package pin to the internal chip will be about 0.05 in and introduce about 0.8 nH. Chip capacitors serving the secondary bypass function typically display similar inductances, independent of the capacitor value. Thus, increasing C_{B2} to accommodate the C_{B1} change does not materially degrade stability.

8.5 Power-Supply Decoupling

Power-supply bypass provides the best first defense against supply-coupled noise and related instability. However, sometimes the bypass that preserves stability does not sufficiently reduce the noise coupled from other circuitry. Then, supply decoupling filters may be required. This typically occurs when the supply-line noise contains frequencies above the amplifier's useful PSRR range. There, the lack of PSRR often couples supply-line noise straight through the amplifier to its output. Subsequent filtering would remove this noise except for the intermodulation effect of this higher-frequency noise. As described before, the filtering effectively removes the direct effect of this higher-frequency error signal without restricting signal bandwidth. However, this filtering cannot similarly remove noise components that intermodulation distortion down shifts into the amplifier's useful frequency range. There, the higher-frequency supply noise reacts with the amplifier's signal and distortion sources, producing error signals in the lower-frequency range of the amplifier.

8.5.1 Decoupling alternatives

Then, prevention supersedes cure in the reduction of supply-coupled noise. Power-supply decoupling filters prevent the higher-frequency signals from even reaching the amplifier. The simplest decoupling alternatives include R-C, L-C, or R-L-C low-pass filters as illustrated in Fig. 8.9. There, three low-pass filter configurations represent decoupling circuits placed between a power-supply source V_+ and the amplifier's positive supply pin V_P. While shown as a single capacitor, the C_B there represents the combined capacitance of all capacitors used for the supply bypass of V_P. In practice, the negative supply line would have an identical filter, but the positive supply case shown here serves to illustrate the decoupling principles.

Each of the three filter alternatives adds a series impedance in the supply line to combine with the bypass capacitance C_B, forming a low-pass filter. Then, the high-frequency noise contained in V_+ drops

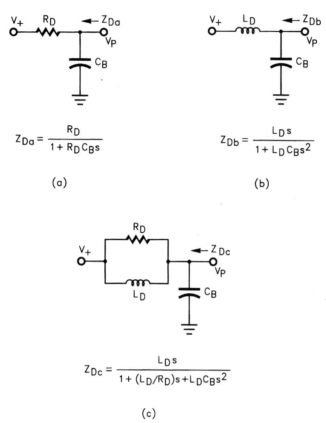

Figure 8.9 Power-supply decoupling introduces series impedance between the V_+ output of the power supply and the op amp's V_P terminal to intercept high-frequency supply transients introduced by other circuitry.

across the series impedance rather than transferring to the V_P supply pin of the op amp. The series impedance added does not have to be mounted close to the amplifier to provide the desired decoupling. However, this impedance should be placed so that it does not conduct the supply-current drain of circuitry other than the amplifier.

The type of impedance added determines the relative cost and decoupling effectiveness. This effectiveness depends both upon the filter response and the line impedance presented to the amplifier. The R-C solution of Fig. 8.9a offers the lowest cost and still produces appreciable decoupling results. With this circuit, decoupling resistance R_D absorbs and dissipates the higher-frequency, transient energy, transferring a more stable voltage to the C_B capacitance. Analysis of the decoupling circuit expresses this action in the classic R-C filter

response
$$V_P = \frac{V_+}{1+R_D C_B s}$$

This response rolls off the supply-line noise with a single-pole at $1/2\pi R_D C_B$.

However, at lower frequencies, this simple R-C decoupling increases the supply-line impedance, potentially increasing the PSRR error. To define this impedance effect, assume that the V_+ power supply presents its ideal zero output impedance, placing R_D effectively in parallel with C_B. This makes the decoupling impedance seen from the V_P terminal

$$Z_{\text{Da}} = \frac{R_D}{1+R_D C_B s}$$

Thus, at low frequencies the R-C filter increases line impedance by the amount R_D. To limit this effect, the R_D resistance must be kept low, and a later discussion describes its value selection.

Alternately, L-C filtering removes this impedance increase and produces a double- rather than single-pole roll off of supply-line noise. However, this alternative introduces another impedance increase through a supply-line resonance. Figure 8.9b illustrates this case with inductor L_D simply replacing the previous resistor R_D. Then, the L-C filter presents the low impedance of the inductor to lower-frequency current demands from the V_+ power supply. The filtering response of this alternative rolls off supply-line noise with a double pole at $f = 1/2\pi \sqrt{L_D C_B}$ as described by

$$V_P = \frac{V_+}{1+L_D C_B s^2}$$

Compared with the R-C filter, the L-C filter reduces the supply-line impedance but only at lower frequencies. The impedance seen from the V_P terminal becomes

$$Z_{\text{Db}} = \frac{L_D s}{1+L_D C_B s^2}$$

At lower frequencies, the s term of the Z_{Db} numerator assures low impedance, but this same term tends to increase Z_{Db} at higher frequencies. Counteracting this increase, the s^2 term of the denominator returns the net impedance response to a single-pole roll off like that attained with the R-C filter.

Thus, the L-C filter offers two-pole roll off of supply-line noise and reduced low-frequency line impedance. However, at an intermediate frequency, the Z_{Db} impedance increases due to a new L-C resonance.

From an energy perspective, the L_D inductor stores rather than dissipates the transient energy of supply-line noise, giving rise to this resonance. At most frequencies, the inductor simply releases its stored energy to the bypass capacitance and the amplifier in a more gradual manner than initially presented by a supply-noise transient. This smoothing of the energy release protects the amplifier from supply-line noise by reducing the associated frequency content to a range where amplifier PSRR remains high. However, at the inevitable L-C resonance the Z_{Db} impedance rises and the filtering fails. As indicated by the Z_{Db} denominator, $1 + L_D C_B s^2$, the L_D-C_B combination produces a double pole and a resonance at $f_{\text{rDb}} = 1/2\pi \sqrt{L_D C_B}$. This resonance increases rather than decreases the line impedance in the vicinity of f_{rDb}.

This resonance makes the simple L-C decoupling filter an uncertain alternative for op amp circuits having high noise gains like the photodiode amplifier. At resonance, energy stored in the inductor by supply-line transients circulates between the capacitor and inductor until dissipated by some energy drain. In the ideal L-C case, this circulation continues indefinitely, sustaining an oscillation signal. In practice, parasitic resistances and the amplifier help absorb that circulating energy, but the amplifier's PSRR error and gain offer replenishment to support this oscillation. There, supply-voltage transients or noise containing frequency components in the range of the f_{rDb} resonance can initiate this condition.

Adding a detuning resistor removes the resonance and the oscillation uncertainty but reduces the noise roll off. Shown in Fig. 8.9c, this alternative combines the R_D and L_D series impedances of the preceding two cases. The combination provides both the energy absorption of the R-C filter and the reduced low-frequency impedance of the L-C filter. For this R-L-C combination the filtering response becomes

$$V_P = \frac{[1 + (L_D/R_D)s]V_+}{1 + (L_D/R_D)s + L_D C_B s^2}$$

Here, the s^2 term of the denominator initially continues the two-pole roll off advantage of the basic L-C case. However, the s term of the numerator indicates a response zero that returns this roll off to single pole. Thus, adding R_D to the L-C filter reduces the filter attenuation.

Still, R_D improves decoupling effectiveness by detuning the line impedance resonance that otherwise adds gain to the parasitic PSRR feedback loop. Analysis of the Z_{Dc} impedance, as seen from the V_P terminal, yields

$$Z_{\text{Dc}} = \frac{L_D s}{1 + (L_D/R_D)s + L_D C_B s^2}$$

This expression repeats that of the previous Z_{Db} except for the $(L_D/R_D)s$ term of the denominator. That term separates the circuit's two poles to detune the characteristic L-C resonance and remove the associated impedance rise.

8.5.2 Selecting the decoupling components

Two compromises guide the selection of the decoupling filter components. The first repeats the bypass capacitance selection described before, optimizing the value of C_B to maintain frequency stability. Design equations developed earlier guide this selection. The second compromise guides the selection of the series impedance to limit noise coupling without significantly degrading other performance.

The series impedance, R_D or the R_D and L_D combination, must be large enough to intercept high-frequency line transients but not so large as to compromise the circuit's lower-frequency PSRR error signal. The power-supply decoupling reduces the supply-line noise produced by other circuitry but can increase the parasitic feedback produced by the amplifier itself. At lower frequencies, the bypass capacitors fail to roll off the line impedance, and adding the series impedance increases the supply-line impedance seen by the amplifier. The discussion of Sec. 8.1 describes the noise-coupling effect of this impedance. The net effect upon circuit performance remains a function of multiple variables and discourages analysis. However, following the previous bypass capacitor selection guidelines and then setting $R_D \sim 5 \,\Omega$ serves most amplifiers that supply output currents in the 10-mA range. This 5-Ω guideline, combined with the 1-Ω guideline for C_B, produces a filter with at least a 6:1 reduction of higher-frequency noise. Also, adding just 5 Ω generally retains lower-frequency line impedance at an acceptable level. Amplifiers supplying greater output current require correspondingly lower values for R_D. For R-C filters, this R_D selection completes the decoupling design.

For R-L-C filters, the L_D selection remains. Adding the L_D inductor of Fig. 8.9c benefits performance as long as the inductance chosen forms a detuned resonant circuit with the C_B and R_D chosen above. Making L_D larger minimizes the line impedance at lower frequencies but potentially increases it at the L-C resonance. For optimum detuning of this resonance, the L_D impedance should equal R_D at the frequency of the potential resonance. From the Fig. 8.9b discussion, the undamped resonance would occur at the frequency $f_{rDb} = 1/2\pi \sqrt{L_D C_B}$. There the impedance of L_D equals $Z_{LD} = 2\pi L_D f_r = \sqrt{L_D/C_B}$, and setting this equal to R_D yields

$$L_D = R_D{}^2 C_B$$

Here, capacitance C_B represents the combined value of the bypass capacitors connected to the associated supply terminal.

References

1. H. Ott, *Noise Reduction Techniques in Electronic Systems*, John Wiley, New York, 1976.
2. J. Graeme, "Feedback Models Reduce Op Amp Circuits to Voltage Dividers," *EDN*, June 20, 1991, p. 139.

Chapter

9

Reducing External Noise Effects

Diminishing returns eventually limit the noise reduction achieved through measures focused upon the photodiode amplifier itself. External noise sources impose a background noise floor that requires attention to the amplifier's environment rather than the amplifier. This background noise typically results from the parasitic noise coupling of external electrostatic and magnetic sources. These sources can dominate noise performance, overriding any effort to further reduce the direct noise effects of the amplifier.

Limiting the effects of these external noise sources requires attention to amplifier location, shielding, the circuit structure, and the physical arrangement of circuit components.[1,2,3] Electrostatic and magnetic noise signals enter the amplifier circuit through parasitic mechanisms. Between these sources and the amplifier, parasitic mutual capacitances couple electrostatic interference, and parasitic mutual inductances couple magnetic interference. For both sources, maximizing the physical separation of the amplifier from the source produces the greatest coupling reduction. Shielding provides the next greatest reduction but with different shield material requirements for electrostatic and magnetic coupling. Reduction of electrostatic coupling simply requires a shield material having high electrical conductivity to shunt capacitively coupled currents to ground. Reduction of magnetic coupling requires a shield material having high magnetic permeability or significant thickness to attenuate the magnetic field that can otherwise couple through a simple conductive shield. Fortunately, the higher frequencies that make magnetic coupling most effective also make simpler shield materials more effective.

Differential circuit structures and loop-minimizing component arrangements provide further reduction of coupled noise. Converting the photodiode amplifier to a differential-input form activates the CMR of the op amp for rejection of both electrostatic and magnetic coupling effects. However, limited amplifier response range limits this solution to lower frequencies. Component arrangements that minimize loop areas reduce magnetic coupling by minimizing the mutual inductances that couple the noise. Similarly, coaxial power-supply and load connections minimize higher current loops that could generate significant magnetic fields.

9.1 Reducing Electrostatic Coupling

Electric field coupling, such as from the power line, supplies noise signals through the mutual capacitances that exist between any two objects. The two objects serve as the capacitor's plates, and the intervening air or other material acts as the capacitor's dielectric layer. AC voltage differences between the objects drive these mutual capacitances, coupling noise currents from one to the other. Electrostatic shielding intercepts this noise current and shunts it to ground. Alternately, differential-input photodiode configurations turn the coupled signal into common-mode signals for removal by the op amp's CMR. In the simplest case, one op amp performs this differential function by utilizing the opposite polarity currents of the photodiode's two terminals. However, this configuration may limit bandwidth. A three-op-amp solution removes this limit and actually increases bandwidth over that achieved with the basic photodiode amplifier.

9.1.1 Electrostatic shielding

Shield material and grounding determine the shield effectiveness. Making the shield with a material of high electrical conductivity ensures that the coupled currents produce little voltage drop across the shield. Then, little of the original field continues within the shield boundaries. To be effective, the shield's ground must be earth ground because only this remains a common reference for the separate objects involved in the coupling. Further, the shield should be connected to the system ground to minimize the effects of the parasitic capacitances introduced. All shield enclosures form parasitic capacitances with each component shielded. When left ungrounded, the shield develops a signal voltage determined primarily by the circuit's larger signals through parasitic capacitance coupling. Then, the shield signal couples back into the circuit through other parasitic capacitances. Returning the shield connection to the system common removes the shield signal and the associated coupling. Then, shield-conducted capacitive currents from the output of a current-to-voltage converter shunt to ground rather than to the circuit's input. This avoids a bandwidth restriction otherwise imposed by the associated parasitic bypass of the circuit's feedback resistor. However, the shield then produces a capacitance from the circuit's input to ground, adding to the noise gain peaking described in Sec. 5.1.2.

9.1.2 The differential-input current-to-voltage converter

Most electrostatic noise coupling results from the ac power line and couples to all points of a circuit nearly equally. The equal coupling makes this interference a natural candidate for removal by CMR, and op amp CMR readily removes signals at the power-line frequency. However, the basic current-to-voltage converter fails to utilize the amplifier's CMR capability. There, the single-ended input connection of the circuit disables this feature. Switching to a differential-input connection activates CMR for reduced noise sensitivity and reduced dc error as well. However, as will be described, this connection does reduce bandwidth. Also, op amp CMR does not totally replace shielding because the electrostatic coupling to the two amplifier inputs will not be exactly the same. Often, CMR serves as a backup to shielding by removing the residual coupling that passes through shield imperfections.

Both the op amp and the photodiode possess differential qualities that combine to reject noise coupling effects. The opposite polarity inputs of the op amp separate any differentially applied signal from common-mode signals. The photodiode produces a differential signal in the form of opposite polarity currents, as seen from the two-diode terminals. Current flowing into one photodiode terminal flows out of the other, producing a differential current signal. Applying this signal to the amplifier's two inputs produces the differential-input current-to-voltage converter of Fig. 9.1. There, a second current-to-voltage

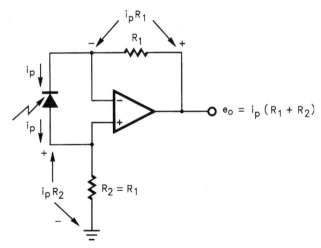

Figure 9.1 Adding a second current-to-voltage conversion resistor R_2 and bootstrapping the photodiode produces a differential-input circuit for rejection of external noise coupling.

conversion resistance R_2 and a bootstrap connection for the photodiode differentiate this circuit from the basic case. These simple differences produce easily anticipated changes in circuit gain and dc offset but less obvious changes in noise performance.

First, the circuit changes increase the current-to-voltage gain and automatically provide for dc offset reduction. In the figure, the added resistor R_2 develops a second signal voltage from the anode current of the diode. This signal $i_p R_2$ adds to the normal signal $i_p R_1$ increasing the output signal to $e_o = i_p(R_1 + R_2)$. Resistor R_2 also occupies the position of the offset compensation resistance described in Sec. 2.2.1. As described there, making the circuit's two resistances equal produces canceling offset effects with the amplifier's two input currents. As will be seen, making $R_2 = R_1$ also balances the circuit for optimum rejection of noise interference. The differential connection here reduces offset error even further through two other features. First, connecting the diode's anode to the top of R_2 bootstraps the diode on that resistor's voltage. Then, the dc compensation voltage developed across R_2 no longer drops across the diode, avoiding the associated diode leakage error. The second feature permits a reduction of resistance values. With equal resistances, this differential connection doubles the current-to-voltage gain through $e_o/i_p = 2R_1$. Then, cutting the resistance values in half maintains the same gain as before and reduces the circuit's offset sensitivity to amplifier input currents.

This differential-input connection also provides the desired CMR of coupled noise as shown in Fig. 9.2. There, setting the photodiode current to zero permits a focus upon the electrostatic coupling effects alone. Electrostatic noise source e_e represents any ac voltage source that creates an electric field in the vicinity of the photodiode amplifier. This source couples noise currents i_{ne} through mutual capacitances C_M to the amplifier's two inputs. It might first seem that the coupling effects to the two would differ because of greatly different impedances at those inputs. At the amplifier's noninverting input, the circuit presents an impedance equal to R_2. In contrast, feedback produces a virtual ground at the amplifier's inverting input. There, feedback absorbs any current injected without producing a corresponding voltage change, indicating zero impedance.

In spite of this impedance difference, feedback also equalizes the coupling effects at the two inputs by equalizing the signal voltages impressed upon the two C_M capacitances. Feedback forces the voltage at the amplifier's inverting input to track that at the noninverting input, removing any signal voltage imbalance across the two C_M components. With equal signal voltages, the two capacitances couple equal i_{ne} noise currents to the circuit's two inputs. There, the two i_{ne}

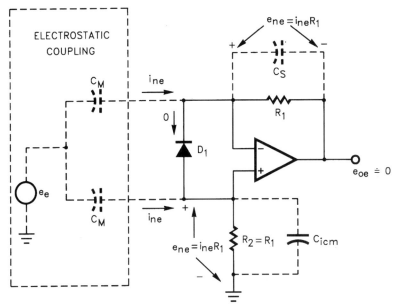

Figure 9.2 The differential input circuit of Fig. 9.1 accepts equal noise currents i_e from an electrostatic source to produce two equal e_{ne} noise effects that the op amp's CMR rejects.

currents develop equal and counteracting $e_{ne} = i_{ne}R_1$ noise voltages on the circuit's two resistors. Making $R_2 = R_1$ as shown ideally produces a zeroed output response, $e_{noe} \approx 0$, from the electrostatic coupling.

The accuracy of the above noise cancellation depends upon the matching of mutual capacitances, resistors, and parasitic capacitances. For a given noise source, locating the two circuit inputs equidistant from the source largely matches the associated mutual capacitances. Then, matching the values for the two circuit resistors assures accurate cancellation of the i_{ne} signal effects, until higher frequencies are reached. There, differences in parasitic capacitive shunting imbalance the net impedances that react with the i_{ne} currents. Typically, only about 0.5 pF of stray capacitance shunts R_1, but the significantly larger, common-mode input capacitance of the op amp shunts R_2. This difference removes the benefit of CMR at higher frequencies, but the rejection typically remains high at the troublesome power-line frequency.

This capacitance imbalance also imposes different bandwidth limits for the circuit's two signal paths. A signal current driving R_2 encounters the shunting of the larger C_{icm} capacitance before C_S significantly shunts the current driving R_1. This difference produces a frequency response characterized by two gain plateaus that are separated by

6 dB. The earlier roll off produced by R_2 and C_{icm} often limits the circuit's useful bandwidth to a frequency significantly lower than that achieved with the basic current-to-voltage converter.

9.1.3 Other noise effects of the differential-input connection

This connection also alters the noise resulting from the photodiode amplifier itself. However, simple adjustments to the earlier Sec. 5.3.1 noise results adapt them to this case. Here, the added resistor and the bootstrap connection alter each of the fundamental E_{noR}, E_{noi}, and E_{noe} noise components of the photodiode amplifier. These components represent the RMS output noise voltages resulting from the circuit's resistance and the amplifier's input noise current and voltage. Figure 9.3 models the differential-input case of Fig. 9.2 with the circuit's noise sources and capacitances separated from the amplifier and the

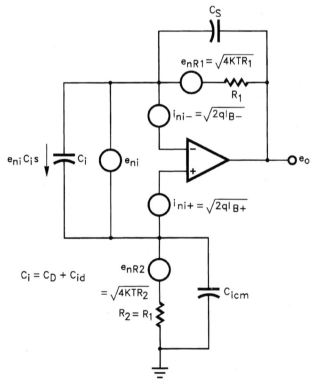

Figure 9.3 The differential-input connection of Fig. 9.2 adds the noise effects of R_2 directly through e_{nR2} and indirectly through the interaction of i_{ni+} and e_{ni} with this resistance.

resistors. There, C_i represents the net capacitance of the input circuit, $C_i = C_D + C_{id}$, where C_D is the diode capacitance, and C_{id} is the differential-input capacitance of the amplifier. Superposition analysis sets the photodiode current to zero in this model to permit a focus on the noise effects alone. The noise sources shown represent spectral noise densities and require separate analyses to determine the RMS output noise resulting from each source.

The addition of R_2 for this differential-input circuit affects noise both directly and indirectly. Directly, the presence of R_2 introduces an additional noise source to E_{noR} and, indirectly, this resistance increases E_{noi} and E_{noe}. First, the noise voltage of R_2 increases the output noise component E_{noR} produced by the circuit resistances. This added noise source, $e_{nR2} = \sqrt{4KTR_2}$, drives the amplifier's noninverting input and the shunting C_{icm} capacitance. Superposition analysis defines the resulting output noise effects by first considering all other signal and noise sources set to zero. Then, at lower frequencies, e_{nR2} effectively drives a voltage follower, transferring directly to the circuit output with a noise gain $A_{nR2} = 1$. At higher frequencies, C_{icm} absorbs the noise energy of R_2, shorting out the e_{nR2} source. The intervening pole at $f_{p2} = 1/2\pi R_2 C_{icm}$ rolls off the corresponding noise gain as expressed by $A_{nR2} = 1/(1 + jf/f_{p2})$. As described in Sec. 5.1.3, applying the RMS noise integral to the product of the e_{nR2} noise density and the A_{nR2} noise gain defines the resulting RMS output noise voltage. Then,

$$E_{noR2}^2 = \int_0^\infty |A_{nR2} e_{nR2}|^2 \, df$$

$$= \int_0^\infty \frac{4KTR_2}{1 + f^2/f_{p2}^2} \, df$$

and $E_{noR2} = \sqrt{2KTR_2 \pi f_{p2}}$, where $f_{p2} = 1/2\pi R_2 C_{icm}$.

The noise density of R_1, e_{noR1}, produces a second RMS resistor noise component. As with the basic photodiode amplifier, the $e_{noR1} = \sqrt{4KTR_1}$ noise also transfers to the circuit output with unity gain until rolled off by a pole at $f_{pf} = 1/2\pi R_1 C_S$. The resulting noise gain, $A_{nR1} = 1/(1 + jf/f_{pf})$, and the noise density e_{noR1} produce an RMS output noise analogous to that above, $E_{noR1} = \sqrt{2KTR_1 \pi f_{pf}}$. An RMS summation of E_{noR1} with E_{noR2} defines the circuit's net output noise E_{noR}, produced by the circuit's resistors. For the typical $R_2 = R_1$ case this combination produces

$$E_{noR} = \sqrt{E_{noR1}^2 + E_{noR2}^2} = \sqrt{2KT \pi R_1 (f_{pf} + f_{p2})}$$

where $f_{pf} = 1/2\pi R_1 C_S$, and $f_{p2} = 1/2\pi R_2 C_{icm}$.

Compared with the basic current-to-voltage converter, where $E_{\text{noR}} = \sqrt{2KT\pi R_f f_{\text{pf}}}$, the above noise component typically decreases. Comparing the $R_1(f_{\text{pf}}+f_{p2})$ and $R_f f_{\text{pf}}$ products of the two E_{noR} equations reveals this decrease. For equal gains in the two cases $R_1 = R_f/2$, making $R_1(f_{\text{pf}}+f_{p2}) = (R_f/2)(f_{\text{pf}}+f_{p2})$. Then, only the condition of $f_{p2} = f_{\text{pf}}$ would make $R_1(f_{\text{pf}}+f_{p2}) = R_f f_{\text{pf}}$ to equalize the two E_{noR} results. However, in almost all cases, $f_{p2} < f_{\text{pf}}$ by a significant amount, making $R_1(f_{\text{pf}}+f_{p2}) < R_f f_{\text{pf}}$ and producing a lower E_{noR} for the differential input case. Note, however, that this noise reduction actually results from the decreased bandwidth f_{p2} of the differential-input circuit.

The addition of R_2 indirectly introduces another noise component through this resistor's reaction with the amplifier's input noise current. With the basic photodiode amplifier, the noise current of the amplifier's inverting input, $i_{\text{ni}-}$ of Fig. 9.3 produces an output noise voltage by flowing through the feedback resistance. That also occurs here and produces a noise voltage with R_1 of $e_{\text{nRi}-} = (i_{\text{ni}-})R_1$, where $i_{\text{ni}-} = \sqrt{2qI_{B-}}$ and I_{B-} is the input bias current of the amplifier's inverting input. Being across R_1, this noise voltage appears in series with e_{nR1}, and the amplifier supplies the same noise gain to the two noise voltages. From before, $A_{\text{nR1}} = 1/(1+jf/f_{\text{pf}})$, and applying the RMS integral to the product of $e_{\text{nRi}-}$ and A_{nR1} yields the RMS output noise voltage produced by $i_{\text{ni}-}$, $E_{\text{noi}-} = R_1 \sqrt{q\pi f_{\text{pf}} I_{B-}}$.

With the differential case here, the noise current of the noninverting input $i_{\text{ni}+}$ also produces an output noise voltage. An analogous analysis defines its effect. Here, $i_{\text{ni}+} = \sqrt{2qI_{B+}}$, where I_{B+} is the input bias current of the amplifier's noninverting input. Noise current $i_{\text{ni}+}$ flows through the R_2 resistance, producing the noise voltage $e_{\text{nRi}+} = (i_{\text{ni}+})R_2$ across R_2. There, $e_{\text{nRi}+}$ appears in series with e_{nR2}, and the previous $A_{\text{nR2}} = 1/(1+jf/f_{p2})$ noise gain applies to $e_{\text{nRi}+}$ as well. This $e_{\text{nRi}+}$, A_{nR2} combination produces an RMS output noise component of $E_{\text{noi}+} = R_2\sqrt{q\pi f_{p2} I_{B+}}$. An RMS combination of $E_{\text{noi}+}$ and $E_{\text{noi}-}$ produces the net RMS output noise due to amplifier input currents. Assuming the typical conditions of $R_2 = R_1$ and $I_{B+} = I_{B-}$,

$$E_{\text{noi}} = R_1 \sqrt{q\pi I_{B-}(f_{\text{pf}}+f_{p2})}$$

where $f_{\text{pf}} = 1/2\pi R_1 C_S$, and $f_{p2} = 1/2\pi R_2 C_{\text{icm}}$. Compared with the basic current-to-voltage converter, where $E_{\text{noi}} = R_f\sqrt{q\pi f_{\text{pf}} I_{B-}}$, the above noise component increases due to the added f_{p2} term but decreases for a gain-equivalent choice of $R_1 = R_f/2$. Typically, the combined increase and decrease effects produces a net decrease in noise but, again, only due to the reduced bandwidth imposed by f_{p2}.

The third and final noise effect of the R_2 addition increases the E_{noe} effect of the amplifier input noise voltage e_{ni}. Evaluation of the net e_{ni} effect requires RMS integration of the product $A_{ne}e_{ni}$, where A_{ne} is the noise-gain response presented by the circuit to e_{ni}. However, the multiple capacitances of the differential-input photodiode amplifier produce a complex A_{ne} response, making direct mathematical analysis cumbersome. For noise analysis purposes, an intuitive examination of the circuit produces a simpler and adequate approximation of the A_{ne} response. For this examination consider all signal sources except for e_{ni} set to zero in Fig. 9.3. Then, only e_{ni} produces an output signal, and at lower frequencies, the circuit capacitances have no effect. For this combination, a loop equation shows that $e_o = e_{ni}$ in response to a low-frequency noise gain of $A_{neL} = 1$. At higher frequencies, the presence of e_{ni} across C_i develops a noise current $e_{ni}C_i s$ that flows through both R_1 and R_2. The resulting noise voltages on these resistors produce an A_{ne} gain rise, marked by a feedback zero at approximately $f_{zf} = 1/2\pi (R_1 + R_2)C_i$. For the typical $R_2 = R_1$ case, this zero becomes $f_{zf} = 1/4\pi R_1 C_i$. At even higher frequencies, C_{icm} and C_S bypass R_2 and R_1, limiting the effect of the noise current developed in C_i. There, the circuit's capacitances control the response to e_{ni}, and a loop equation defines a high-frequency noise gain for e_{ni} approximated by

$$A_{neH} = 1 + C_i \frac{C_{icm} + C_S}{C_{icm}C_S}$$

Figure 9.4 displays the resulting $e_{noe} = A_{ne}e_{ni}$ output noise-density response. This result closely resembles the original e_{noe} response of Sec. 5.2.3, making the earlier results applicable here with just a few modifications. As before, five regions separate the $A_{ne}e_{ni}$ response of the figure to simplify an otherwise complex integration. Here, changes in f_{zf}, f_i and the shape of region 3 response differentiate this case from the earlier result. In regions 1 and 2 of Fig. 9.4, the low-frequency noise gain $A_{neL} = 1$ makes the e_{noe} response again follow the e_{ni} noise response of the amplifier. Then in region 3, the e_{noe} response rises due to the A_{ne} response zero at $f_{zf} = 1/4\pi R_1 C_i$, as identified above. The resulting response rise continues with a constant slope in region 3 until temporarily interrupted by the C_{icm} bypass of R_2. This bypass cuts the noise gain in half and introduces a dip in the curve. However, this dip occurs late in the region 3 span, and ignoring it produces little error. Thus, assuming a straight line response between f_{zf} and f_{pf} adequately approximates the region 3 response and extends previous results to this case.

The C_{icm} capacitance also affects regions 4 and 5 of the curve where the intercept frequency f_i defines a boundary for the two. This frequency marks the crossing of the circuit's $1/\beta$ curve and the A_{OL} response. There, $f_i = \beta_H f_c$, where β_H is the high-frequency value of

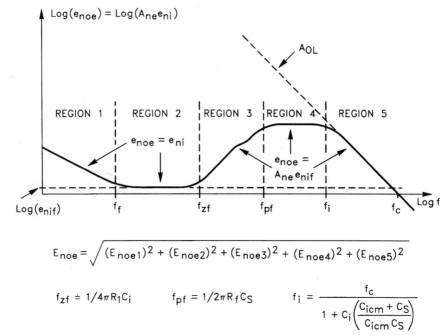

Figure 9.4 The differential-input model of Fig. 9.3 modifies the frequency response of the photodiode amplifier's output noise density with a negligible dip in region 3 and a change in the f_i boundary between regions 4 and 5.

feedback factor and f_c is the amplifier's unity-gain crossover frequency. The factor β_H follows from the Fig. 9.3 derivation of the high-frequency noise gain A_{neH} through the relationship $\beta_H = 1/A_{neH}$. Then, substituting the previous A_{neH} result in $f_i = \beta_H f_c = f_c/A_{neH}$ produces

$$f_i = \frac{f_c}{1 + C_i(C_{icm} + C_S)/C_{icm}C_S}$$

Making the preceding changes to E_{noR}, E_{noi}, f_{zf}, and f_i adapts the original noise analysis of Sec. 5.3.1 to this differential-input case. This analysis covers only the noise produced by the photodiode amplifier components and not the noise introduced by external sources. Then, the analysis of self-induced noise for the Fig. 9.2 follows the steps listed below:

$E_{noe1} = e_{nif}\sqrt{f_f \ln(f_f/f_1)}$, where e_{nif} is the e_{ni} noise floor level and f_f is the $1/f$ noise corner frequency. As explained in Sec. 5.2.3, setting $f_1 \approx 0.01$ Hz avoids a theoretical contradiction and ensures analytical accuracy.

$E_{\text{noe2}} = e_{\text{nif}}\sqrt{f_{\text{zf}} - f_f}$, where $f_{\text{zf}} = 1/4\pi R_1 C_i$, and $C_i = C_D + C_{\text{id}}$. Here and in the E_{noe4} analysis, numerical evaluation can produce the square root of a negative number. Then, substitute zero for the result as explained in Sec. 5.3.1.

$E_{\text{noe3}} = (e_{\text{nif}}/f_{\text{zf}})\sqrt{\left(f_{\text{pf}}^3 - f_{\text{zf}}^3\right)/3}$, where $f_{\text{pf}} = 1/2\pi R_f C_S$, and C_S is the stray feedback capacitance.

$E_{\text{noe4}} = (1 + C_i/C_S)e_{\text{nif}}\sqrt{f_i - f_{\text{pf}}}$, where $f_i = f_c/[1 + C_i(C_{\text{icm}} + C_S)/C_{\text{icm}}C_S]$.

$E_{\text{noe5}} = (e_{\text{nif}} f_c)\sqrt{1/f_i}$, where f_c is the op amp's unity-gain crossover frequency.

$E_{\text{noe}} = \sqrt{E_{\text{noe1}}^2 + E_{\text{noe2}}^2 + E_{\text{noe3}}^2 + E_{\text{noe4}}^2 + E_{\text{noe5}}^2}$, the total RMS output noise resulting from the spectral noise voltage e_{ni} of the amplifier input.

$E_{\text{noR}} = \sqrt{2KT\pi R_1(f_{\text{pf}} + f_{p2})}$, where $f_{\text{pf}} = 1/2\pi R_1 C_S$, and $f_{p2} = 1/2\pi R_2 C_{\text{icm}}$.

$E_{\text{noi}} = R_1\sqrt{q\pi I_{B-}(f_{\text{pf}} + f_{p2})}$.

$E_{\text{no}} = \sqrt{E_{\text{noR}}^2 + E_{\text{noi}}^2 + E_{\text{noe}}^2}$.

9.1.4 An alternate differential-input photodiode amplifier

The preceding photodiode amplifier reduces sensitivity to electrostatic coupling but sacrifices signal bandwidth. There, the C_{icm} bypass of R_2 imposes a new bandwidth limit at a frequency lower than the normal limit imposed by the C_S bypass of R_1. An alternative circuit restores bandwidth by removing signal swing from C_{icm}. Shown in Fig. 9.5, this differential-input circuit connects the photodiode between the inputs of two current-to-voltage converters and subtracts the two converter output signals with a difference amplifier. The resulting circuit becomes the common three-op-amp instrumentation amplifier with the photodiode replacing the normal gain-setting resistor. With this connection, grounded noninverting inputs for A_1 and A_2 assure zero signal swing across the op amps' C_{icm} capacitances to avoid the preceding bandwidth limit. Then, the bandwidth limit returns to the stray capacitance bypass of the converter feedback resistors R_1 and R_2. Making $R_2 = R_1$ sets this limit at $1/2\pi R_1 C_S$ for both converters and balances the circuit for optimum rejection of electrostatic coupling. Compared with the basic photodiode amplifier, bandwidth actually improves here due to the dual gain setting resistors. For a given gain

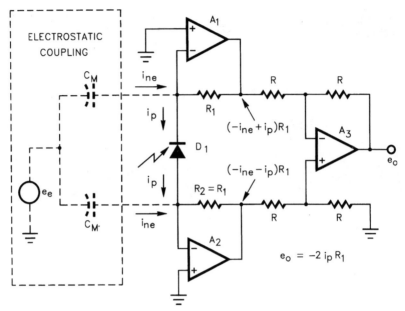

Figure 9.5 Connecting the photodiode to a grounded-input instrumentation amplifier removes the common-mode swing of Fig. 9.1, avoiding the bandwidth limit imposed there by common-mode input capacitance.

level, dual resistors permit a 2:1 resistance reduction, reducing bandwidth sensitivity to stray capacitance by the same factor.

As with the preceding circuit, the differential inputs of this alternative circuit separate the electrostatic interference for rejection by CMR while amplifying the desired signal. Mutual capacitances C_M again couple equal electrostatic interference currents i_{ne} to R_1 and R_2. For $R_2 = R_1$, the equal currents produce equal and in-phase voltages $-i_{ne}R_1$ at the outputs of A_1 and A_2. This makes the $-i_{ne}R_1$ signals a common-mode signal at the input of the A_3 difference amplifier. From there, the subtracting action of this amplifier rejects these identical signals. In contrast, the differential nature of the photodiode current i_p produces opposite-phase voltages at the outputs of A_1 and A_2. This current flows into one diode terminal and out of the other, producing i_pR_1 at the output of A_1 and $-i_pR_1$ at the output of A_2. Then, the subtracting action of the A_3 difference amplifier delivers an output voltage of $e_o = -2i_pR_1$. This separation of differential and common-mode signals depends upon the CMR of A_3, and this rejection decreases with increasing frequency. However, the rejection generally remains high at the power-line frequency most often involved in electrostatic noise coupling.

9.1.5 Other noise effects of the differential-input alternative

While it does reduce electrostatic noise, the Fig. 9.5 alternative adds the noise effects of several circuit components. Instead of one amplifier and one resistor, this alternative includes three amplifiers and six resistors. Fortunately, A_3 and the four resistors of its difference amplifier enter the signal path after the signal gain provided by A_1 and A_2. This placement makes the noise effects negligible for the five components of the A_3 difference amplifier. Then, the noise introduced by A_1, A_2, R_1, and R_2 represents the noise contribution of the photodiode amplifier. These circuit components form two matching current-to-voltage converters that each introduce the noise components previously summarized in Sec. 5.3.1. Simple RMS addition of the two converter's noise effects produces the net result for this case. For matched converter components, this addition of two equal noise components, $\sqrt{E_n^2 + E_n^2}$, simply produces $E_n\sqrt{2}$, introducing a $\sqrt{2}$ multiplier to the circuit's noise components. Adding this multiplier to the results of the earlier Sec. 5.3.1 results adapts those results to this case. This addition modifies E_{noR}, E_{noi}, and E_{noe1} through E_{noe5}, which represent the noise effects of the circuit's resistance noise, current noise, and the five subcomponents of voltage noise. Making these changes to the Sec. 5.3.1 summary produces the equivalent noise analysis summary for the differential-input alternative of Fig. 9.5.

$E_{\text{noe1}} = e_{\text{nif}}\sqrt{2f_f \ln(f_f/f_1)}$, where e_{nif} is the e_{ni} noise floor level, and f_f is the $1/f$ noise corner frequency. As explained in Sec. 5.2.3, setting $f_1 \approx 0.01$ Hz avoids a theoretical contradiction and ensures analytical accuracy.

$E_{\text{noe2}} = e_{\text{nif}}\sqrt{2(f_{\text{zf}} - f_f)}$, where $f_{\text{zf}} = 1/2\pi R_1(C_i + C_S)$, and $C_i = C_D + C_{\text{id}} + C_{\text{icm}}$. For applications requiring feedback compensation C_f, replace C_S with $C_f + C_S$ here and in the steps that follow. Here and in the E_{noe4} analysis, numerical evaluation can produce the square root of a negative number. Then, substitute zero for the result as explained in Sec. 5.3.1.

$E_{\text{noe3}} = (e_{\text{nif}}/f_{\text{zf}})\sqrt{2(f_{\text{pf}}^3 - f_{\text{zf}}^3)/3}$, where $f_{\text{pf}} = 1/2\pi R_1 C_S$, and C_S is the stray feedback capacitance.

$E_{\text{noe4}} = (1 + C_i/C_S)e_{\text{nif}}\sqrt{2(f_i - f_{\text{pf}})}$, where $f_i = f_c C_S/(C_i + C_S)$.

$E_{\text{noe5}} = (e_{\text{nif}} f_c)\sqrt{2/f_i}$, where f_c is the op amp's unity-gain crossover frequency.

$E_{\text{noe}} = \sqrt{E_{\text{noe1}}^2 + E_{\text{noe2}}^2 + E_{\text{noe3}}^2 + E_{\text{noe4}}^2 + E_{\text{noe5}}^2}$, the total RMS output noise resulting from the spectral noise voltage e_{ni} of the amplifier inputs.

$E_{\text{noR}} = 2\sqrt{KTR_1 \pi \text{BW}_t}$, where $\text{BW}_t = 1/2\pi R_1 C_S$. For applications that add a feedback capacitance C_f, replace C_S with $C_f + C_S$.

$E_{\text{noi}} = R_1 \sqrt{2q \pi \text{BW}_t I_{B-}}$.

$E_{\text{no}} = \sqrt{E_{\text{noR}}^2 + E_{\text{noi}}^2 + E_{\text{noe}}^2}$.

9.2 Reducing Magnetic and RFI Coupling

Magnetic noise coupling and radio-frequency interference (RFI) introduce circuit noise through a common coupling mechanism, mutual inductance. There, the interference source acts like the primary of a transformer, and circuit loops resemble secondary windings. While often considered separately, RFI simply represents the higher-frequency form of parasitic magnetic coupling. However, in shielding, this frequency distinction suggests the separation. The frequency of the magnetic source greatly influences the effectiveness of magnetic shielding materials. At lower frequencies, only the magnetic property of ferrous metals permits practical shield thickness. However, at higher frequencies, decreases in both the magnetic response and in the shield thickness requirement make copper a good alternative material. There, even the copper layer of a ground plain becomes effective. In addition to shielding, careful choosing of the amplifier's physical and electrical configurations reduces magnetic coupling effects. Minimizing the physical areas of circuit loops reduces the coupling efficiency, and using a differential-input structure makes CMR reduce the residual effect.

9.2.1 Magnetic shielding

A comparison of magnetic coupling with electrostatic coupling and a transformer analogy illustrate the added shielding requirement of the magnetic case. The electrostatic shielding described before simply requires a shield material of high electrical conductivity. There, the high conductivity shorts the currents transmitted through mutual capacitances to ground. As long as these currents develop little voltage across the shield, no significant field continues within the shield boundaries. However, magnetic fields couple through mutual inductances, rather than capacitances, and produce voltage rather than current signals. The coupled signals develop in all circuit loops within the field and have relative magnitudes determined by loop areas. For the magnetic field, a high-conductivity electrostatic shield only forces an equipotential condition at the shield boundaries. Grounding this shield does not terminate the magnetic field and may just simply establish a zero voltage reference. By analogy, grounding the center tap of a transformer's secondary winding establishes a reference voltage but does not terminate the transformer's coupling. Thus, an electrostatic shield distorts the magnetic field but does not necessarily remove the field energy. Some portion of the field's energy continues the field within the shield boundaries.

Fortunately for noise reduction, some of the magnetic field energy dissipates in eddy currents and ohmic absorption within the shield. Shield material selection optimizes this dissipation with different results for low and high frequencies. To effectively reduce magnetic coupling, the shield must absorb the field energy as the field travels through the shield's walls. A combination of high conductivity and high magnetic permeability produces the greatest field absorption. High electrical conductivity ensures that the induced shield currents produce little voltage drop across the shield, preventing continuation of the field through electrostatic coupling within. High magnetic permeability ensures efficient absorption of the magnetic field moving through the shield. At the power-line frequency, common to the most serious magnetic coupling effects, the magnetic permeability of ferrous materials reduces magnetic coupling an order of magnitude better than other shield materials.[2] For such materials, the realignment of magnetic dipoles in the ferrous material consumes magnetic field energy in the form of eddy currents. This improvement occurs even though the electrical conductivity of steel is about an order of magnitude lower than that of copper. Also, at power-line frequencies, the shield thickness required for a copper shield becomes prohibitive due to the great skin depth of this material.

At radio frequencies (RF), copper becomes a more reasonable, but never superior, alternative shield material. There, both the magnetic permeability of ferrous metals and the skin depths of all metals drop dramatically. Above 10 kHz, the magnetic permeability of ferrous metals drops due to the finite time required to realign the material's magnetic dipoles. The shorter periods of high-frequency signals preclude the realignment and the conversion of field energy to eddy currents. Then, a shield must remove magnetic field energy through ohmic absorption as reflected by skin depth. Skin depth indicates the thickness of a given shield material required to attenuate a magnetic field by a factor of $e = 2.73$. Fortunately, skin depth also decreases at higher frequencies, due to the shorter wavelength of the signal. Thus, at high frequencies, the decreased permeability of ferrous materials reduces their relative advantage over copper, and the reduced skin depths make the required copper thickness more practical. However, for a given shield thickness, steel still retains about a factor of 3 absorption advantage.

9.2.2 Circuit reductions of magnetic coupling

As described above, physically separating and shielding the photodiode amplifier from a magnetic noise source offers the best protection against magnetic noise coupling. However, the amplifier's physical and electrical configurations also offer powerful alternatives. Configuring the amplifier layout for minimum physical loop areas and designing for CMR also reduce magnetic noise coupling. Minimizing the areas of the circuit's loops minimizes the mutual inductances that couple the magnetic noise signal. Careful component layout achieves this by placing the photodiode amplifier's components close to the op amp. Simply minimizing the lengths of component interconnections generally produces the desired area reductions. Finally, for lower-frequency noise, an op amp's CMR reduces coupling effects when using a differential-input configuration for the photodiode amplifier. Then, by matching loop areas and distances from a noise source the amplifier's CMR rejects the resulting, equal noise coupling effects.

For photodiode amplifiers, the most confusing task in magnetic coupling reduction can be the identification of the receptor loops. The physical arrangement of the circuit's components form these loops in several ways, and Fig. 9.6 shows the loops produced by the differential-input photodiode amplifier of Fig. 9.1. Here, three loops illustrate those commonly formed with photodiode amplifiers through the circuit connections of the photodiode, the feedback, and the load. The op amp would seem to break these loops, but as will be seen, the amplifier's feedback action continues them.

Differing amplifier responses and a grounding connection differentiate the three loops shown. First, the photodiode's connection to the op amp inputs forms loop L_3 that develops magnetic coupling signal e_{m3}. In this loop, it would seem that the amplifier's very high input impedance would break the loop by interrupting an otherwise continuous conductive path. However, feedback forces the voltage between the amplifier inputs to zero, just as if a short-circuit connected them. This completes the L_3 loop in spite of the high input impedance. The equivalent short circuit of this case parallels the condition of a shorted transformer secondary. There, the transformer continues its magnetic coupling and simply delivers its short-circuit output current. Here, the L_3 loop acts as the receptor or secondary side of a parasitic transformer and similarly delivers its short-circuit current i_{ss}. This current can-

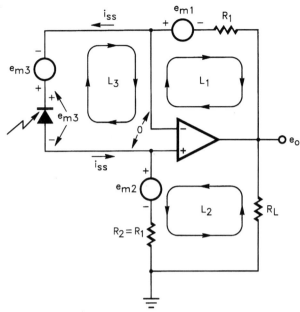

Figure 9.6 Photodiode amplifiers form magnetic receptor loops, in spite of the high-impedance interruption of the op amp inputs, due to the amplifier's output response to signals at its inputs.

not flow into the amplifier inputs because of their high impedances. Instead, feedback supplies this current through R_1 and forces the return current into R_2. To do so, the amplifier develops an output voltage supporting the resulting voltage drops on the two resistors. This condition maintains the necessary zero voltage between the amplifier inputs and retains a short-circuit appearance to the L_3 loop.

The above feedback action and the photodiode capacitance make the L_3 loop a serious noise source for the photodiode amplifier. The output voltage developed by feedback above is actually an amplified replica of the e_{m3} noise. To determine the corresponding noise gain, let superposition set e_{m1} and e_{m2} to zero and consider i_{ss} to be only that current produced by e_{m3}. This current develops a voltage across the photodiode that counteracts e_{m3} in the L_3 loop and retains zero voltage between the op amp inputs. Generally small, e_{m3} does not produce a forward bias condition for the photodiode but keeps the diode junction near its off state. At lower frequencies, the off impedance of this junction equals the photodiode's very large dark resistance, R_D. Together, e_{m3} and R_D produce a very small current of $i_{ss} = e_{m3}/R_D$. The circuit conducts this current from R_1 and to R_2, producing an output voltage of $e_o = i_{ss}(R_1 + R_2) = 2i_{ss}R_1$. Combining the last two equations shows the low-frequency noise gain to be $A_{nL} = e_o/e_{m3} = 2R_1/R_D$. Typically, the very large value of R_D makes this gain less than unity, and the amplifier actually attenuates low-frequency e_{m3} signals. However, at higher frequencies, the diode's capacitance C_D shunts the diode impedance, replacing R_D with $1/C_D s$ in the noise gain equation. Then, $A_{nH} = 2R_1 C_D s$, and this noise gain displays a response zero at $1/4\pi R_1 C_D$. For large values of R_1 and C_D, this zero makes the high-frequency noise gain rise into the hundreds much like the noise gain peaking described in Sec. 5.1.2. This noise gain makes the L_3 loop particularly troublesome for photodiode amplifiers.

Next, consider the L_2 loop and its coupled signal e_{m2}. Less obvious, this loop results from connecting R_2 and R_L together at the circuit's ground return and from a signal path through the amplifier. The ground connection here does not disable the loop reception, just as a grounded tap on a transformer secondary does not disable the transformer's magnetic coupling. Nor does the amplifier interrupt the loop, as seen by a superposition analysis with $e_{m1} = e_{m3} = i_{ss} = 0$. Then, e_{m2} drives the input of a voltage follower, and feedback makes the amplifier output follow e_{m2}, just as if a short circuit connected e_{m2} to R_L. This equivalent short circuit completes the L_2 loop through the amplifier. Due to the follower response of this case, the circuit presents a noise gain of $A_n = 1$ to e_{m2}.

The L_2 loop demonstrates another benefit of the differential-input photodiode amplifier of the figure. There, the circuit's bootstrap connection isolates the diode capacitance from the e_{m2} signal, reducing noise gain. Returning the anode of the diode to the amplifier's noninverting input bootstraps the diode on the e_{m2} signal at that input. In contrast, common photodiode amplifiers connect the anode to ground, impressing e_{m2} across the diode. Then, e_{m2} reacts with the diode capacitance, increasing high-frequency noise gain, just as e_{m3} did before.

The final circuit loop L_1 depends upon the equivalent short-circuits described for L_2 and L_3. As described, feedback action produces these equivalents from the amplifier's noninverting input to both the inverting input and the output. Visual examination of Fig. 9.3 shows that making the two equivalent connections completes the L_1 loop. Superposition analysis again determines the corresponding noise gain but using $e_{m2} = e_{m3} = i_{ss} = 0$ in this case. Under these conditions, one side of the e_{m1} source connects to the virtual ground that feedback produces at the amplifier's inverting input. The other side connects to the output through R_1 and, with $i_{ss} = 0$, no current flows in this resistor. Thus, $e_o = -e_{m1}$ for a noise gain of $A_n = -1$.

Minimizing the preceding loops reduces but does not eliminate the noise from magnetic coupling. Some finite loop areas and coupling always remain. However, the differential-input photodiode amplifier offers further noise reduction through op amp CMR. The circuit of Fig. 9.6 rejects magnetic coupling much as it did electrostatic coupling in Fig. 9.1. Previously, matching R_1 and R_2 resistances activated the amplifier's CMR. Here, matching the L_1 and L_2 loops produces similar results. These loops produce the e_{m1} and e_{m2} signals at opposite amplifier inputs and superposition analysis with $e_{m3} = i_{ss} = 0$ defines an output voltage of $e_o = e_{m2} - e_{m1}$. Thus, making $e_{m2} = e_{m1}$ produces noise cancellation of these two effects at the circuit output.

This requires matching the physical areas of the L_1 and L_2 loops, their distances from an interfering magnetic source, and their orientations relative to that source. Matching these three features equalizes e_{m1} and e_{m2}, making their net effect a common-mode signal at the amplifier's inputs. Matching loop areas and distances equalizes the magnitudes of e_{m2} and e_{m1}, and the matching distances also produces a first-order phase equalization. Accurate phase matching, as required for high common-mode cancellation, also requires matched loop orientations relative to the magnetic source. Then, the subtraction performed by the amplifier's differential inputs produces optimum rejection of magnetic noise coupling. Most often, this noise reduction opportunity aids in the rejection of low-frequency local noise sources such as power transformer interference. At higher frequencies, the amplifier's CMR declines, reducing its benefit to noise reduction.

There, RFI filtering somewhat replaces CMR to counteract the accompanying increase in magnetic coupling efficiency. This increase potentially dominates noise performance, often due to the RFI of coresident digital circuitry. However, the bandwidth limitation of the amplifier permits a filtering solution. The limited bandwidth of the op amp typically restricts the photodiode amplifier's response to frequencies well below the RF range. This permits filter removal of RFI signals at the amplifier output without restricting the circuit's useful bandwidth.

Still, output filtering does not necessarily remove the effects of higher-level RFI. The input circuit of an op amp can act like an RF detector, separating a lower frequency envelope from a carrier.[4] There, larger RF signals drive the emitter-base junctions of bipolar input transistors to produce a rectifying action. Then, the transistor's junction capacitances store the envelope level of the RF signal at the amplifier input. If not disabled by this interference, the amplifier transmits an amplified replica of the envelope to the circuit output. FET input op amps significantly reduce the likelihood of this effect due to the larger voltage swing required to produce a rectifying action. Fortunately, the lower input bias currents of FET amplifiers also make them the predominate choice for photodiode applications.

9.3 Reducing Multiple Coupled-Noise Effects

Coupled noise seldom provides clear clues to its origin, and successful reduction of this noise requires a modified, troubleshooting approach. In a given application, any one of the preceding noise reduction techniques may offer the best solution, or the application may require a combination of two or more. Without definitive clues, trial-and-error application of the noise reduction practices prevails. However, the combinational possibility requires a modification to this approach. Traditionally, when a given trial implementation fails to solve the problem, you remove this implementation and proceed to the next. However, with the potential for combinational effects, effective noise coupling reduction requires a delay in this removal. When the trial of a first noise reduction technique fails to solve the problem, do not remove it before applying the next technique. The effect of the first may be masked by the requirement for a second, and removing the first at this point may suboptimize the final solution. Instead, proceed through the series of noise reduction practices described above, leaving all implementations in place until the end. Then, with the coupling reduction optimized, begin to remove those implementations which originally appeared to have no effect. In the reduced noise background of the optimized condition, the actual noise effects of the individual implementations become apparent. Remove the ineffective ones at this point but not before.

9.4 Minimizing Magnetic Field Generation

The preceding discussions focus upon the photodiode amplifier's sensitivity to coupled noise, overlooking this amplifier's potential for producing such noise. Simple current flow in a conductor produces a magnetic field capable of coupling noise back into the amplifier or to other circuitry. Current flows associated with a photodiode amplifier typically lack the magnitude needed to produce a significant field, except at the amplifier output. There, the load current supplied by the amplifier potentially produces a magnetic noise source, but attention to coaxial returns minimizes this potential.

As illustrated in Fig. 9.7, the photodiode amplifier supplies the load current i_L to load resistance R_L, often producing a significant current flow in connecting conductors. Closer examination of this current's path reveals current flow in a field-generating loop. For the

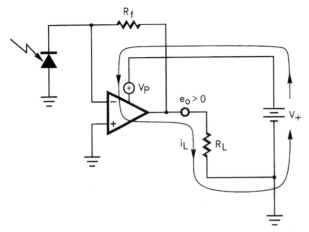

Figure 9.7 Load current supplied by the amplifier actually flows in a loop area capable of producing magnetic interference.

case shown, the amplifier produces a positive output voltage e_o by drawing the load current i_L from the V_+ positive supply and delivering it to R_L. Then, R_L's ground connection returns the current to V_+, completing the loop. A similar current conduction loop results for negative e_o values through the amplifier's negative power supply. In each case, the magnitude and frequency content of i_L and the physical area of the conduction loop determine the strength of the magnetic field generated. Application requirements dictate the characteristics of i_L, but physical layout options permit minimizing the loop area to greatly reduce the resulting field strength.

Coaxial conduction of the supply and return of the i_L current minimizes this loop area. For the simplest illustration of this principle, Fig. 9.8 shows this current conduction with a coaxial cable. Other, more practical options generally suffice, and later discussion presents these options using the principles illustrated here. In the figure, the cable's center conductor supplies the current to the amplifier, and the cable shield returns it to the power-supply common. Just the appearance of the resulting figure communicates the reduced area of the conduction loop, supplying an intuitive sign of magnetic field reduction. In practice, the relative area reduction far exceeds that presented by the short cable length of the illustration.

The significance of this area reduction becomes apparent when considering the magnetic fields produced by the supply and return conductors. To minimize loop area, the supply and return conductors must follow essentially the same physical path. These conductors carry the i_L current in opposite directions along this common path,

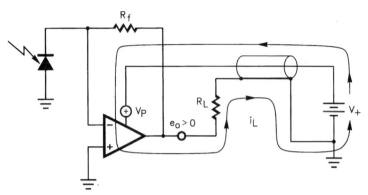

Figure 9.8 Coaxial return of the load current minimizes loop area and the associated magnetic interference.

producing equal magnetic fields of opposite orientation. These fields cancel, neutralizing the net magnetic field. The coaxial cable optimizes this canceling effect by producing a common centroid or axis for the two current flows. While difficult to illustrate, the return current conducted in the cable shield of Fig. 9.8 distributes throughout the shield, enveloping the center conductor. Averaged around the circular shield, the net magnetic field produced has a centroid coincident with the center conductor. Thus, the supply and return conductors of the coaxial cable permit precise magnetic field neutralization. However, coaxial cable connections represent an impractical solution to the multitude of current conduction loops found on a typical circuit board.

Fortunately, adjacent board traces or a ground plane provide nearly coaxial performance for the practical case. Adjacent circuit board traces illustrate this solution, but the ground plane best supplies it. Conceptually adjacent traces, supplying and returning a given current, approximate the coaxial condition. This configuration again minimizes loop area and only slightly compromises the ideal coaxial conduction of the cable above. With adjacent traces, the space separating the two differentiates the central axes of the two magnetic fields produced, slightly reducing the magnetic field cancellation. Still, the adjacent trace solution also remains impractical for all but the simplest of circuit boards. Layout implementation of a return trace adjacent to each signal trace soon becomes a product of conflicting demands.

In practice, a ground plane avoids the adjacent trace complication and provides similar coaxial performance. There, serendipity assures the coaxial return at the higher frequencies where magnetic coupling becomes increasingly significant.[5] A return current delivered to the ground plane follows the path of least impedance back to the power-supply common. At lower frequencies, the ground plane resistance controls this impedance, and the return current typically follows the shortest path to the supply common. This does not establish a coaxial return, but the signal's lower frequency prevents the production of a noise-significant magnetic field. At higher frequencies, the ground plane inductance controls this return impedance and automatically assures a coaxial return. There, the loop area of the conduction path corresponds to inductance, and the loop's magnetic field controls the direction of current flow. Seeking the path of least impedance, the return current follows a path paralleling that of the supply current, minimizing the inductance and the associated magnetic field. The physical separation between the supply trace and the ground plane still differentiates the axes of the two conduction paths but only by the thickness of the circuit board.

References

1. R. Morrison, *Grounding and Shielding Techniques in Instrumentation*, 2nd ed., John Wiley, New York, 1977.
2. H. Ott, *Noise Reduction Techniques in Electronic Systems*, John Wiley, New York, 1976.
3. J. Graeme, "Systematic Approach Makes Op-Amp Circuits Resist Radiated Noise," *EDN*, July 20, 1995, p. 93.
4. Y. Sutu and J. Whalen, "Statistics for Demodulation RFI in Operational Amplifiers," *IEEE International Symposium on Electromagnetic Compatibility*, August 23, 1983.
5. P. Browkaw and J. Barrow, "Grounding for Low- and High-frequency Circuits," *Analog Dialog*, 23-3, 1989.

Chapter

10

Position-Sensing Photodiode Amplifiers

Photodiodes sense the position of a light beam through the diodes' photo responses, but several variables potentially degrade sensing accuracy. These variables introduce offset and gain errors in the light-to-voltage conversion process. A single photodiode provides some measure of a light beam's position through the magnitude of the diode's output signal. However, background light conditions also influence this magnitude and could require calibrated and often impractical measurement conditions. Fortunately, background light only adds an offset to the diode output, and differential measurement removes its effect. Adding a second, matching photodiode and monitoring the difference between the two diode outputs removes the equal offsets produced by the two. The differential-input nature of op amps readily extracts this difference signal through several circuit configurations.[1] There, two photodiodes provide one axis of position information, and replicating the two-diode solution provides multiaxis information.

Still, these position-sensing solutions remain vulnerable to variations in the light source intensity and the photodiode's responsivity. Both variables affect the gain of the position measurement, and simple differential monitoring does not remove gain error effects. Once again, calibrated measurement conditions can temporarily remove these effects, but light sources age and the photodiode responsivity displays a strong temperature sensitivity. Signal normalization provides a more permanent solution by dividing the diodes' difference signal by their mean signal. There, gain variations affect the mean and difference signals equally, making their normalized quotient immune to gain changes. An analog divider extracts this normalized signal with the greatest accuracy, but analog multiplier replacements for the divider simplify the circuitry. Alternately, linear photodiode arrays remove the need for normalization by providing a digital indication of position. There, the diodes of the array need only indicate which diode resides closest to the incident beam as resolved by evaluating their relative rather than absolute signal magnitudes.

10.1 Direct-Displacement Monitor Amplifiers

In the simplest case, two photodiodes directly indicate a beam position through the difference in their output signals. Four circuits below produce the desired difference signal with varying degrees of circuit complexity and performance. Using three op amps, a differential-photodiode amplifier provides the best overall performance and offers simplified noise filtering. Alternately, connecting the photodiodes differentially reduces the amplifier requirement to the basic current-to-voltage converter. However, this simpler circuit compromises accuracy, offset, and noise. Modifying this simpler circuit bootstraps the photodiodes to restore offset and noise performance, but this modification potentially reduces bandwidth. These first three circuits provide single-axis monitoring with two photodiodes. A fourth circuit extends the monitoring to two axes using four photodiodes.

10.1.1 Single-axis monitoring with a differential-photodiode amplifier

While somewhat complex, the differential-photodiode amplifier of Fig. 10.1 develops the photodiode difference signal with the greatest accuracy. This circuit first converts the two photodiode currents into voltages and then subtracts those voltages. In the figure, A_1 and A_2 perform as traditional current-to-voltage converters, and A_3 performs as a difference amplifier. Fundamentally, the circuit becomes the common three-op-amp instrumentation amplifier[2] with the gain set resistor removed and photodiodes connected to the gain set points. The output voltages of A_1 and A_2 contain both difference and common-mode signals. A_3 separates the two, delivering the difference signal to the circuit output as $e_o = (i_{p1} - i_{p2})R_f$.

This difference signal reflects the displacement information desired through the difference in the two photodiode currents. Note that equal illumination of the two photodiodes makes $i_{p1} = i_{p2}$ and $e_o = 0$, indicating a beam centered between the two diodes. Similarly, greater illumination of D_1 makes $i_{p1} > i_{p2}$ and $e_o > 0$, indicating a displacement in the D_1 direction, and the converse conditions result for greater illumination of D_2. Currents i_{p1} and i_{p2} respond linearly to illumi-

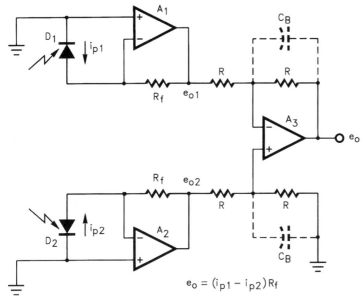

Figure 10.1 The differential-photodiode amplifier resolves the position of a light source relative to two photodiodes as represented by the difference in diode output currents.

nation intensity and make the magnitude of e_o a direct measure of displacement.

Evaluation of the e_o equation confirms the circuit's removal of background light source effects. A photodiode produces an output current of $i_p = r_\phi \phi_e$, where r_ϕ is the diode's flux responsivity and ϕ_e is the radiant flux in watts.[3] The combination of signal and background light sources makes $\phi_e = \phi_{es} + \phi_{eB}$ and $i_p = r_\phi(\phi_{es} + \phi_{eB})$. Applying this i_p expression to $e_o = (i_{p1} - i_{p2})R_f$ produces $e_o = [(r_{\phi1}\phi_{es1} - r_{\phi2}\phi_{es2}) + (r_{\phi1} - r_{\phi2})\phi_{eB}]R_f$. Here, the term $(r_{\phi1} - r_{\phi2})\phi_{eB}R_f$ describes the residual offset produced by the background flux ϕ_{eB} that illuminates the two diodes equally. Simply matching the photodiode responsivities, $r_{\phi1} = r_{\phi2}$, reduces this offset to zero, leaving $e_o = (r_{\phi1}\phi_{es1} - r_{\phi2}\phi_{es2})R_f$.

Circuit mismatches produce error in the actual e_o signal, but resistor adjustments produce compensating effects. Specifically, the two photodiodes will not have identical responsivities, and the high value of the R_f resistors makes their matching difficult. Both effects produce gain imbalance in the circuit's two signal channels, and this disturbs the circuit's difference function. Adjusting the R_f resistors would compensate for both effects, but again, the high resistance value makes such adjustment difficult. Instead, adjusting the more reasonable R resistors of the difference amplifier produces a compensating gain imbalance. Also, the Fig. 10.1 configuration facilitates lower initial mismatch through the use of monolithic dual photodiodes. Such duals typically have common cathodes as accommodated by the grounded cathode configuration of the differential amplifier.

Dual op amps also benefit this amplifier configuration by reducing offset voltage. For a current-to-voltage converter, an offset error results from the flow of op amp input current into the high feedback resistance. Normally, adding a matching resistance in series with the op amp's noninverting input produces a compensating offset voltage. However, that compensating voltage drops across the photodiode, producing a leakage current that often overrides the compensation resistor's benefit. The circuit here removes the need for that resistor by automatically providing offset voltage compensation through CMR. There, placing the matched amplifiers of a dual in the A_1 and A_2 positions delivers matched amplifier input currents to the two R_f resistors. The resulting, equal offsets voltages produced at the outputs of A_1 and A_2 present a common-mode signal for rejection by the A_3 difference amplifier.

10.1.2 Noise reduction with the differential-photodiode amplifier

This amplifier permits reducing the noise effects of both internal and external noise sources. Simplified filtering reduces the effects of the amplifier's noise sources, and CMR reduces the effects of noise coupled from external sources. For a different application, Secs. 9.1.4 and 9.1.5 develop detailed noise analyses for both noise effects of the differential amplifier, and those analyses apply to this application as well.

The simplified filtering opportunity results from a difference in signal and noise bandwidths. As described in Chap. 6, the current-to-voltage converter often delivers a greater bandwidth to internal noise sources than to the signal, but filtering can restrict both bandwidths to the same frequency and remove noise without affecting signal. The previous discussion also presents several such filtering methods for the basic current-to-voltage converter. Any of these methods could be applied to the two current-to-voltage converters of this circuit; however, this circuit's third op amp offers a simpler solution. As shown with dashed lines, simple capacitance bypass of two resistors of the difference amplifier produces the desired filtering. This adapts the difference amplifier to also serve as a low-pass filter and avoids the added complexity of previous solutions.

Such bypass can disturb the circuit balance that produces the difference amplifier's CMR, and ideally, matching bypass responses are required. However, two factors make this matching less critical here. First, the dominant noise sources of the circuit do not present a common-mode signal to the difference amplifier. There, the output noise signals of the two current-to-voltage converters result from different amplifiers and remain uncorrelated rather than common. Thus, the subtraction operation of the A_3 difference amplifier cannot make the two noise effects cancel, and the circuit's CMR offers no benefit here. Instead, the uncorrelated noise components of e_{o1} and e_{o2}, e_{no1} and e_{no2}, pass through the difference amplifier with unity gain and are added in RMS fashion to produce a final output noise component

$$e_{no3} = \sqrt{e_{no1}^2 + e_{no2}^2}$$

Introducing the C_B capacitors produces a pole in the unity-gain responses to e_{no1} and e_{no2}, making

$$e_{no3} = \frac{\sqrt{e_{no1}^2 + e_{no2}^2}}{1 + RC_B s}$$

A second factor reducing the bypass balance requirement results from the frequency of the bypass. Setting the bypass break frequency at the signal bandwidth limit moves the primary effect of the bypass to frequencies beyond the range of interest. Still, the circuit depends upon CMR for accurate development of the difference signal, and the effect of bypass imbalance should be evaluated when using this option.

Note that another obvious filtering option does not yield an equivalent result. Capacitive bypass of the R_f conversion resistance would also reduce noise bandwidth. However, as described in Sec. 6.1.1, this bypass simultaneously reduces signal bandwidth by the same factor, leaving the ratio of the two bandwidths unchanged. Then, signal bandwidth declines, and noise still retains the bandwidth advantage.

The differential-photodiode amplifier also aids in the reduction of noise coupling from external sources. Electrostatic and magnetic coupling can be significant noise contributors especially because of the small signals typically developed by photodiodes. The differential nature of this circuit converts those coupling effects into common-mode signals for rejection by the difference amplifier. Chapter 9 describes this noise reduction in detail. Note that this reduction depends upon the common-mode nature of the coupled signals and the circuit's CMR. Thus, the filtering described above potentially compromises the reduction of coupled noise effects as well.

10.1.3 Connecting the photodiodes differentially

With their voltages held to zero by op amp inputs, photodiodes become two-terminal current sources, and current signals sum or subtract easily. This permits a simpler monitor circuit but introduces some performance compromises. Simply connecting two photodiodes in parallel and with opposite orientations in Fig. 10.2 produces the differential output current $i_{p1} - i_{p2}$. Common-mode current merely circulates in the diode loop resulting in no signal output. Then, the simple current-to-voltage converter again serves as the photodiode amplifier. This circuit produces the same $e_o = (i_{p1} - i_{p2})R_f$ as the preceding differential-photodiode amplifier.

However, this simpler circuit also degrades response accuracy, offset, and the rejection of coupled noise. Response accuracy degrades due to the different cathode connections of the diodes. Previously, the grounded cathodes of the two photodiodes permitted the use of monolithic devices that ensure matched responsivities for the two. Offset degrades because the offset compensation resistor R_C now develops a voltage across the photodiodes. That resistor produces a dc offset voltage with the amplifier's noninverting input current to compensate for the voltage developed by the inverting input current with the feedback resistor. However, the voltage on the compensation resistor also appears across the photodiodes, inducing diode leakage currents. These leakages flow through the feedback resistor and produce an output offset. Previously, the differential amplifier held both diode voltages at zero and achieved offset compensation through CMR without the need for compensation resistors. Rejection of coupled noise also degrades, again due to the lack of CMR that the previous circuit used to remove this noise effect.

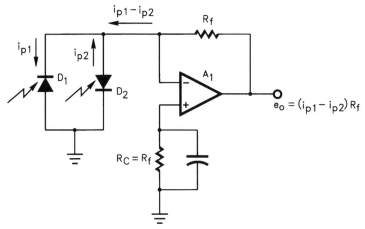

Figure 10.2 Connecting two photodiodes in parallel and with opposite orientations automatically produces a differential signal, reducing the photodiode amplifier to the basic current-to-voltage converter.

10.1.4 An alternate differential-photodiode connection

A simple modification to the Fig. 10.2 circuit restores the previous offset and noise rejection performance at the expense of bandwidth. Shown in Fig. 10.3, the photodiodes now return to the op amp's noninverting input rather than ground. Previously, Sec. 9.1.2 introduced this circuit for a single-diode application. The accompanying noise analysis of Sec. 9.1.3 applies to this case as well. Here, the modified diode return again bootstraps the diodes upon the voltage developed across the compensation resistor R_C. Now, the diodes connect directly across the zero voltage of the op amp's differential inputs, removing diode leakage and the associated offset result. This modified circuit also removes the Fig. 10.2 bypass capacitor that rolled off the noise of R_C there. Then, R_C participates in the current-to-voltage conversion and restores CMR. The $i_{p1} - i_{p2}$ return current of the diodes now flows through R_C, producing a second component of output signal e_o. For its compensation purpose, $R_C = R_f$, and the added e_o component doubles the circuit's current-to-voltage gain. Also, the balanced impedances now connected to the op amp's inputs restore CMR of coupled noise. Electrostatic noise currents coupled to the amplifier's two inputs produce equal and canceling voltages on the circuit's two resistors. Section 9.1.2 describes this noise cancellation in detail.

However, this alternate differential-photodiode circuit potentially reduces bandwidth. Two factors affect bandwidth here, with the first

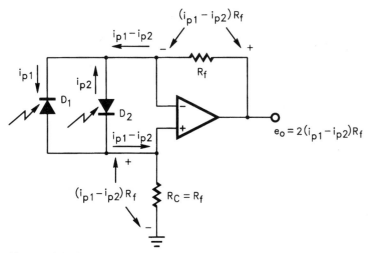

Figure 10.3 Bootstrapping the photodiodes removes Fig. 10.2's major performance compromises but potentially reduces bandwidth.

producing an increase but with the second producing a greater decrease. Bandwidth tends to increase due to the doubled current-to-voltage gain $2R_f$ described above. This permits reducing R_f a factor of 2 to produce the same gain as the preceding circuits. Then, parasitic capacitance bypass of the R_f resistances presents a bandwidth limit at twice the frequency experienced before. However, this alternative connection also adds another bypass capacitance across the compensation resistance as shown in Fig. 10.4. There, the inevitable C_S stray capacitances bypass both resistors, but the op amp adds its common-mode input capacitance C_{icm} to the bypass of R_C. Typically, $C_{icm} \approx 6C_S$, and the combined bypass of R_C rolls off its gain contribution at about one-seventh the frequency of the C_S bypass of R_f. The latter bypass often determines the bandwidth of the photodiode amplifier, so the C_{icm} effect here potentially reduces bandwidth a factor of 7. Combined, the two bandwidth effects potentially produce about a 3.5:1 bandwidth reduction for the alternative differential-photodiode circuit.

The actual frequency response of this circuit exhibits two gain plateaus instead of the customary single plateau. For the typical $R_C = R_f$ case, Fig. 10.5 illustrates this response starting with a low-frequency I-to-V gain plateau at $2R_f$. At an intermediate frequency,

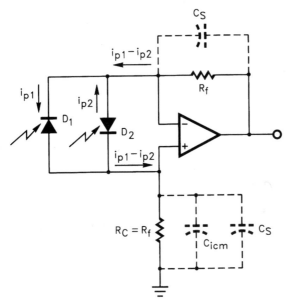

Figure 10.4 The bootstrap connection of Fig. 10.3 makes the amplifier's common-mode input capacitance part of the bandwidth-limiting bypass of current-to-voltage conversion resistance.

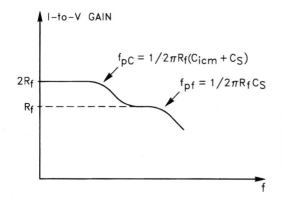

Figure 10.5 The added C_{icm} bypass of Fig. 10.4 introduces a second response pole at f_{pC} that potentially dominates the circuit's bandwidth.

C_{icm} shunts $R_C = R_f$ and produces a response pole at $f_{pC} = 1/2\pi R_f (C_{icm} + C_S)$. This shunt totally removes the gain contribution of R_C but that of R_f remains, interrupting the first gain roll off with a second plateau at a level equal to R_f. At a higher frequency, the stray C_S shunts R_f, producing the second roll off at $f_{pf} = 1/2\pi R_f C_S$. Technically, the second plateau extends the frequency response, but this extension typically serves little practical purpose. For most applications, the roll off at f_{pC} marks the useful bandwidth limit as set by resistor bypass.

However, this limit alone does not necessarily define the final bandwidth. As described in Chap. 3, other limits potentially set the actual bandwidth of the photodiode amplifier. A second limit occurs at the $1/\beta$ intercept with the A_{OL} response at the frequency f_i. Then, f_i competes with f_{pC} in setting the bandwidth, and different circuit conditions make one or the other prevail. High-gain photodiode amplifiers employ very large R_f resistance values that make $f_{pC} < f_i$, and f_{pC} dominates the bandwidth limitations. Then, the alternate differential-photodiode circuit above reduces bandwidth as described. However, lower-gain applications make $f_{pC} > f_i$, removing the significance of resistor bypass from the bandwidth result. Then, this alternate circuit provides the performance advantages described without penalty.

10.1.5 Two-axis monitoring

For multiple-axis position sensing, two photodiodes sense along the line of each axis. Any of the preceding single-axis monitoring circuits of this section serve this purpose with the single-axis circuit simply replicated for each new axis. Figure 10.6 demonstrates this extension for two-axis monitoring using the Fig. 10.1 differential-photodiode amplifier. There, the graphical representation of the diodes depicts their physical placements along X and Y axes. Two diode pairs develop independent output signals representing position relative to the two axes, $e_{ox} = (i_{x1} - i_{x2})R_f$ and $e_{oy} = (i_{y1} - i_{y2})R_f$. These signals serve as indicators of X and Y coordinates or directly effect servo position control.

Two-axis monitoring changes the photodiode performance requirements in two ways. Matching requirements increase, but for some applications, responsivity tolerance requirements relax. As before, the two diodes of an axis pair must match to produce an accurate displacement indication in the e_o difference signal of the axis. However, two-axis monitoring also requires matching one pair to the other for vector angle measurement. Resolving this angle depends upon the ratio of the X and Y signals, and interpair mismatch produces angular error. However, this ratio remains immune to the highly variable diode responsivity, potentially relaxing a requirement there. Responsivity variations due to manufacturing tolerances and temperature changes do not disturb this ratio as long as all diodes experience the same variation. This eases the diode requirements for applications requiring only angular information.

Monolithic photodiode arrays best serve the more stringent matching requirement of this case. As described before, the Fig. 10.1 differential-photodiode amplifier facilitates its use through the grounded connection of the diode cathodes. With both cathodes connected to the same point, that circuit accommodates the common-cathode structure of monolithic photodiodes. Using the Fig. 10.1 amplifier, Fig. 10.6 ex-

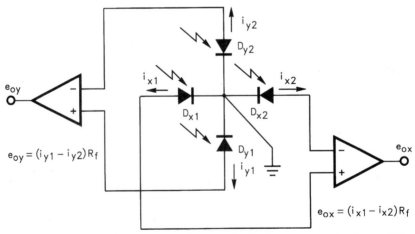

Figure 10.6 Four photodiodes provide two-axis displacement monitoring, and the differential-photodiode amplifier permits the common-cathode connection of a monolithic array.

tends this feature to four photodiodes as shown. Any number of axes can be instrumented in this way with all photodiodes on the same monolithic chip.

10.2 Normalized Monitor Amplifiers

For many applications, the preceding circuits require signal normalization to provide an accurate indication of position. These circuits produce output signals of the form $e_o = (i_{p1} - i_{p2})R_f$, where a difference in diode output signals should directly reflect a light beam's position. However, two other variables typically affect this difference signal by introducing gain error in the light-to-voltage conversion process. Variations in both the light intensity received and the photodiode responsivity to light also affect the magnitudes of the i_{p1} and i_{p2} currents. Received light intensity varies due to changes in the light source and in the medium separating the source from the diodes. Photodiode responsivity varies with manufacturing tolerances and temperature changes. The resulting magnitude variations of i_{p1} and i_{p2} also alter the $i_{p1} - i_{p2}$ difference of the e_o equation, confounding the position sensing. This potentially reduces the e_o difference signal to a relative rather than absolute indication, where e_o only reflects which diode is closest to the source.

To quantify these effects, consider the $e_o = (i_{p1} - i_{p2})R_f$ equation expressed in terms of the light intensity and responsivity variables. The Sec. 10.1.1 discussion reduced this equation to $e_o = (r_{\phi 1}\phi_{es1} - r_{\phi 2}\phi_{es2})R_f$, where r_ϕ represents photodiode diode flux responsivity and ϕ_e represents radiant flux. Consider the case of matched photodiodes where $r_{\phi 1} = r_{\phi 2} = r_\phi$, making $e_o = (\phi_{es1} - \phi_{es2})r_\phi R_f$. Clearly, a variation in responsivity r_ϕ alters gain and changes e_o. Similarly, a variation in light source intensity alters ϕ_{es1} and ϕ_{es2} by the same proportion to also change gain. Thus, both variables potentially make e_o an inaccurate indicator of absolute position. Calibration can remove the effects of these extra variables, but light source aging and diode responsivity changes with temperature make calibration a temporary condition. Some tracking applications only require relative position information because servo adjustments only need to reduce the difference signal to a null. However, the added variables in this signal then require slowing the servo response to prevent the overshoot and ringing otherwise produced by gain increases.

Alternately, signal normalization automatically removes the effects of the added variables to reestablish absolute rather than relative position indication. Normalizing the difference signal against the mean signal,[4] $(i_{p1} - i_{p2})/(i_{p1} + i_{p2})$, divides the difference signal by a signal having equal sensitivity to the added variables. Then, variations in the light intensity or photodiode responsivity change the i_{p1} and i_{p2} variables in both the numerator and denominator of the normalized expression for a net zero result. For example, doubling the responsivity of the diodes replaces an original i_{p1} and i_{p2} with $2i_{p1}$ and $2i_{p2}$, and the change produces canceling effects in the normalized expression above. Three circuits below produce the normalized output with differing degrees of precision and circuit complexity. The first provides the best performance using four op amps and a dedicated analog divider. Next, substituting the less accurate but more common multiplier for the divider eases component selection without increasing circuit complexity. Finally, capitalizing upon circuit summing capabilities reduces the circuit to two op amps and a multiplier. However, this simpler circuit further compromises performance.

10.2.1 Normalizing the differential-photodiode amplifier

The straightforward and most accurate approach to normalization combines the differential-photodiode amplifier with an analog divider as in Fig. 10.7. There, A_1 through A_3 and their associated resistors form the differential amplifier discussed with Fig. 10.1, and A_4 serves as an inverting summer. Together, the amplifier and summer drive the divider's N and D numerator and denominator inputs, and the divider responds with an output response of $10N/D$. The differential amplifier supplies the required difference signal $(i_{p1} - i_{p2})R_f$ directly from the output of A_3. To derive the mean signal, the summer adds and inverts the signals at the outputs of A_1 and A_2, producing $(i_{p1} + i_{p2})R_f$. Then, the divider produces the normalized output

$$e_o = 10 \frac{(i_{p1} - i_{p2})}{(i_{p1} + i_{p2})}$$

As an additional benefit, that the divider introduces a gain of 10 to this result.

Evaluation of the e_o equation confirms the circuit's removal of the effects of variations in light source and photodiode responsivity. As described in Sec. 10.1.1, a photodiode produces an output current of $i_p = r_\phi \phi_e$, where r_ϕ is the diode's flux responsivity and ϕ_e is the radiant flux. Applying this i_p expression to the above e_o produces

$$e_0 = 10 \frac{(r_{\phi 1}\phi_{e1} - r_{\phi 2}\phi_{e2})}{(r_{\phi 1}\phi_{e1} + r_{\phi 2}\phi_{e2})}$$

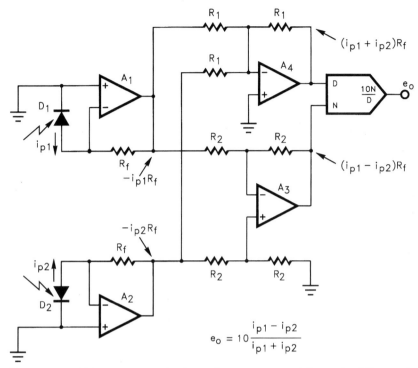

Figure 10.7 Adding an analog divider and a summing amplifier to the differential-photodiode amplifier produces a differential output signal normalized against the mean signal.

Then, assuming responsivity matching for the two diodes, $r_{\phi 1} = r_{\phi 2}$, removes the effects of responsivity manufacturing tolerance and temperature variation in

$$e_o = 10 \frac{(\phi_{e1} - \phi_{e2})}{(\phi_{e1} + \phi_{e2})}$$

There, any change in light source intensity alters ϕ_{e1} and ϕ_{e2} by the same proportion and adds the same multiplier factor to the numerator and denominator of the above expression for a net zero change. For example, doubling the light intensity replaces ϕ_{e1} and ϕ_{e2} with $2\phi_{e1}$ and $2\phi_{e2}$ but leaves e_o above unchanged.

However, the divider added for normalization degrades the circuit's linearity, noise, and offset performance, restricting the dynamic range of the normalized response. Typically, the divider's high noise and offset become dominate error sources, and their effects increase as the denominator signal magnitude decreases. There, a denominator decrease raises the divider's gain of $10N/D$, and the divider's input noise and offset receive greater amplification. This effect typically limits the divider's useful dynamic range to around 40:1. Maintaining a higher-level mean signal $(i_{p1} + i_{p2})R_f$ minimizes the performance degradation.

10.2.2 Easing the normalization

Replacing the analog divider with a multiplier equivalent eases the normalization through expanded selection options. Analog multipliers appear in far greater variety and typically cost less than dividers. In addition, making this replacement in the preceding circuit does not increase overall circuit complexity. However, a divider-connected multiplier does produce greater errors than a dedicated analog divider.

Multipliers perform the division function when connected in the feedback loop of an op amp[3] as in Fig. 10.8. There, the feedback inverts the function of the multiplier as revealed by separate analyses of the circuit's response to the N and D input signals. First, consider a fixed D signal and examine the effect of the N signal. The N signal develops a current N/R_1 at the op amp input. In response, the amplifier produces output signal e_o to drive the multiplier's Y input and force that device to supply the N/R_1 current through R_2. Then, increasing the signal N increases e_o to similarly increase the multiplier output signal e_{om}. This relationship between N and e_o describes that of a numerator response. Next, consider a fixed N signal and examine the effect of the D signal. A fixed N signal requires a fixed e_{om} output from the multiplier as seen from the preceding discussion. In response to the D signal, the multiplier produces an output voltage

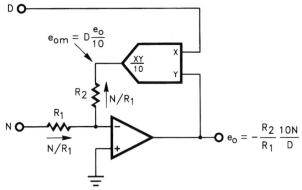

Figure 10.8 Connecting an analog multiplier in the feedback loop of an op amp produces a divider response for replacement of the Fig. 10.7 divider with a more commonly available component.

$e_{om} = D(e_o/10)$, and the multiplier acts as a voltage-controlled amplifier with an input signal of $e_o/10$ and a gain controlled by D. Then, increasing the D signal decreases the e_o signal required to supply a fixed e_{om}. This relationship between D and e_o describes that of a denominator response. Combining the two signal responses described quantifies the net circuit response. From the figure, $e_{om} = D(e_o/10)$ must supply the current N/R_1 through R_2 with the current polarity shown. Then, $e_{om}/R_2 = D(e_o/10R_2) = -N/R_1$, and solving for e_o yields

$$e_o = -\frac{R_2}{R_1}\frac{10N}{D}$$

Successful implementation of this equivalent circuit requires attention to the relative bandwidths of the op amp and the multiplier. Each produces a response roll off, and the two appear in a common feedback loop. The combined roll off potentially produces oscillation unless one of the two bandwidths significantly exceeds the other.[5] Here, stability conditions vary with signal due to the voltage-controlled amplifier function of the multiplier. As described above, feedback gain varies with the signal at the circuit's D input since that multiplies the signal fed back around the amplifier from e_o as indicated by the $e_{om} = D(e_o/10)$ expression. When the D signal reaches its full-scale level, typically 10 V, the gain to the e_o signal reaches its maximum of unity and frequency stability reaches its greatest vulnerability. Stability should be evaluated under this condition. Typically, multipliers achieve greater bandwidths than op amps, and choosing a multiplier with a bandwidth well beyond the f_c crossover of the op amp assures a single, dominant pole for all levels of feedback gain provided by the multiplier.

Above, the divider-connected multiplier produces the response $e_o = -(R_2/R_1)10N/D$. Note the gain inversion provided by the minus sign and the added degree of gain control provided by the R_2/R_1 ratio. Both features simplify the divider replacement in the normalized differential-photodiode amplifier. As described above, the multiplier equivalent of the analog divider requires placing the multiplier in the feedback loop of an op amp. This would seem to add an op amp to a circuit, but the added gain inversion and gain control avoid this for the differential amplifier.

Shown in Fig. 10.9, the modified circuit still uses just four op amps because the circuit no longer requires the previous inverting summing amplifier. Previously, that amplifier developed the mean signal $(i_{p1} + i_{p2})R_f$ from the outputs of A_1 and A_2 and supplied the divider's denominator input. Here, the combination of the multiplier A_4 and the R_3, $2R_3$ resistances form the multiplier equivalent of the previous

236 Chapter Ten

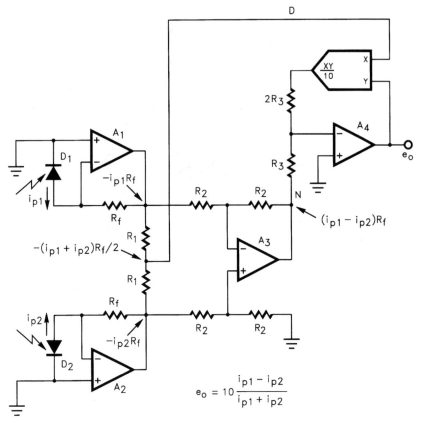

Figure 10.9 Replacing the Fig. 10.7 analog divider with a multiplier adds the A_4 feedback amplifier but eliminates the previous summing amplifier, leaving circuit complexity unchanged.

divider and summer. The equivalent denominator input D senses the center tap of the R_1 voltage divider and that tap produces the signal $-(i_{p1} + i_{p2})R_f/2$. This signal differs from the desired mean through a minus sign and a factor of 2 divisor. However, the gain inversion of the multiplier equivalent neutralizes the minus sign, and the R_3, $2R_3$ resistor network cancels the factor of 2. As a result, the Fig. 10.9 circuit repeats the previous result and delivers the normalized output

$$e_o = 10\frac{(i_{p1} - i_{p2})}{(i_{p1} + i_{p2})}$$

However, the factor of 2 divisor of $D = -(i_{p1} + i_{p2})R_f/2$ does decrease the denominator signal, increasing noise and offset effects as described at the beginning of this section.

10.2.3 Simplifying the normalized differential-photodiode amplifier

Two circuit summation opportunities greatly simplify this amplifier at the expense of some performance characteristics. Shown in Fig. 10.10, this option requires only two op amps instead of the four of the preceding circuits. Both photodiodes again connect directly across the inputs of op amps for the preferred zero diode bias. This circuit also retains canceling effects for dc offsets and coupled noise through the matching R_f conversion resistors connected to the two amplifier inputs. There, amplifier input currents and electrostatically coupled noise currents produce canceling error effects. However, the photodiodes no longer share the common-cathode connection that previously permitted use of matched, monolithic dual devices. Also, as will be seen, this circuit further reduces the D denominator signal, degrading noise and offset performance even more.

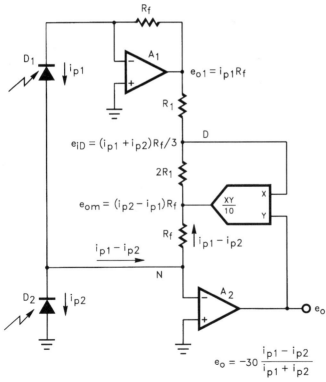

Figure 10.10 Connecting the photodiodes together and adding the R_1, $2R_1$ voltage divider eliminates two of the circuit's four op amps through unique summing opportunities.

Circuit combinations simplify the summations required to generate the difference and mean signals required for normalization. Simply a diode interconnection and a voltage divider provide these summations. In the figure, the multiplier A_2 and the lower R_f resistor again produce the divider function but without the customary numerator input resistor. That resistor would normally be connected between the inverting input of A_2 and the circuit's difference signal voltage. However, current drive removes the need for that resistor here. Connecting the anode of D_1 to the cathode of D_2 results in current summation at their junction. The resulting current, $i_{p1} - i_{p2}$, supplies the numerator input N with the difference signal.

Next, the circuit's mean signal must be developed to drive the denominator input D. The $R_1, 2R_1$ voltage divider derives this signal by summing components of the multiplier and A_1 output signals e_{om} and e_{o1}. The multiplier output signal results from the flow of the $i_{p1} - i_{p2}$ current in the lower R_f resistor and $e_{om} = (i_{p2} - i_{p1})R_f$. Amplifier A_1 performs as a current-to-voltage converter monitoring the current of D_1 and produces $e_{o1} = i_{p1}R_f$. From these two signals, the voltage divider produces the denominator input signal $e_{iD} = 2e_{o1}/3 + e_{om}/3 = (i_{p1} + i_{p2})R_f/3$. Here, the 3 divisor makes the e_{iD} signal depart from the actual mean signal. For the normalization function here, this just introduces a scaling factor, and the $(i_{p1} + i_{p2})$ term of e_{iD} still contains the mean signal information. The 3 denominator term does, however, indicate an attenuated denominator signal, and Sec. 10.2.2 describes how this increases noise and offset errors.

The above signal conditions define the final circuit output voltage through the multiplier's response. As before, the multiplier's $XY/10$ response produces $e_{om} = e_{iD}e_o/10$. Equating this to the previous $e_{om} = (i_{p2} - i_{p1})R_f$ and substituting $e_{iD} = (i_{p1} + i_{p2})R_f/3$ yields the final normalized response for Fig. 10.10

$$e_o = -30\frac{(i_{p1} - i_{p2})}{(i_{p1} + i_{p2})}$$

Here, a -30 factor replaces the previous 10, and the minus sign reflects the inherent phase inversion of the divider-connected multiplier circuit. The increased magnitude reflects the added gain required from the multiplier by the attenuated denominator signal.

Note that the presence of the voltage divider in this circuit introduces the potential for a latch condition. This voltage divider supplies the e_{iD} denominator signal, and that signal must not become a negative voltage for feedback polarity reasons. Positive e_{iD} values assure a positive gain through the multiplier's $e_{om} = e_{iD}e_o/10$ response, and this retains a net negative feedback around the op amp. However, negative values of e_{iD} reverse the polarity of the multiplier's gain, producing positive feedback and the latch condition. In normal operation, the $e_{iD} = (i_{p1} + i_{p2})R_f/3$ assures positive voltages because i_{p1} and i_{p2} can only be positive values. The positive voltage delivered to the voltage divider by A_1 and the voltage divider's ratio assure this normal condition. However, transient conditions during turn on potentially produce a momentary negative voltage at the D input. This does not significantly affect the preceding normalized circuits because the low impedances of op amp outputs drive that input and restore the D signal after the transient. However, in this case, the resistance of the voltage divider may permit a condition in which a negative multiplier output voltage sustains a negative e_{iD} voltage. Should this occur, capacitively coupling an initialization voltage from the positive power supply to the D input prevents the latch condition.

10.3 A Digital Alternative to Normalization

The analog normalization applied in the preceding section removes the effects of variations in light source intensity and photodiode responsivity. Without normalization, those variables reduce the position-sensing ability of a diode pair to a relative rather than absolute measure. Then, a diode pair can only distinguish which diode resides closer relative to an incoming light beam. This relative indication expresses a digital condition that diode arrays exploit to produce an absolute position indication without the need for normalization. There, many photodiodes in a linear array detect position just by identifying which diode resides closest to the received light beam. Such arrays contain photodiodes fabricated side by side in a strip and detect received light differences along the length of the strip. As long as the diode responsivities match, the diode producing the largest output defines the position of the light source relative to the array. For these arrays, monolithic construction assures this matching.

10.3.1 Processing the array signals

To identify the closest diode, a maximum selector circuit[6] compares the output signals of the photodiodes and identifies the largest signal of the group. Normally, maximum selector circuits deliver this signal as the final output, but that is not informative here. Only the identity of the photodiode producing this signal communicates the position information. Modifying the traditional maximum selector produces this indication as shown in Fig. 10.11. There, separate monitor amplifiers receive the output voltages of the individual photodiodes, and the

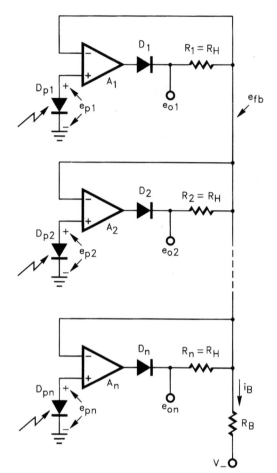

Figure 10.11 A modified maximum selector monitors a photodiode array, producing a series of digital output signals that indicate a light beam's position on the array by the diode producing the high output state.

amplifiers control separate output signals, labeled e_{o1} through e_{on}. All of the photodiodes have grounded cathodes, and this common return permits the use of the common-cathode configuration characteristic of monolithic arrays.

Also, all of the op amps have their inverting inputs connected to a common feedback-sensing line. This common connection produces the signal magnitude discrimination. To this line, each amplifier acts as a voltage follower that would make the line's signal e_{fb} track that amplifier's photodiode voltage. Each amplifier would require this to maintain zero voltage between its own inputs. However, the photodiode voltages will differ, and only one amplifier can achieve this feedback control, as determined by the signal diodes connected in series with the amplifier outputs. Those diodes form a gating network that lets the largest amplifier output take control. Bias resistor R_B connects to the negative supply and would pull e_{fb} to that level except for the counteracting i_B current supplied by an amplifier output. Any of the amplifiers can supply such current to raise the e_{fb} voltage but none can reduce it. To raise this voltage an amplifier supplies a positive i_B current to R_B through the corresponding output diode. Reducing e_{fb} from this level would require a negative amplifier output current and the output diodes D_1 through D_n block such currents. Thus, the amplifier demanding the highest e_{fb} level prevails and sets the voltage of the feedback sensing line. That makes e_{fb} greater than the levels that would satisfy feedback for the other amplifiers. Then, the high open-loop gains of those amplifiers drive their outputs negative, reverse biasing the corresponding output diodes. Then, only the dominant amplifier retains a positive output voltage.

10.3.2 Defining the logic output levels

For this digital application, the circuit's other components translate the amplifier output signals into digital signal levels at the e_{o1} through e_{on} output terminals. These terminals assume low and high digital output levels as determined by a diode voltage drop and a voltage divider. Figure 10.12 illustrates these levels with a two-amplifier simplification of the circuit. For this example, assume $e_{p2} > e_{p1}$ as shown. Then, amplifier A_2 prevails in the control of e_{fb}, producing $e_{fb} = e_{p2}$. This makes $e_{fb} > e_{p1}$ and produces the voltage $e_{p2} - e_{p1}$ between the inputs of A_1. A_1 responds, driving its output voltage negative and reverse biasing D_1. With D_1 in the off state, A_1 operates open-loop, and the high open-loop gain drives this amplifier's output to its negative saturation level near V_-. Also with D_1 off, the current in R_1 is zero, and $e_{o1} = e_{p2} \approx 0.6$ V, representing a low logic level. In contrast, the prevailing amplifier A_2 supplies the current $I_B = (e_{p2} - V_-)/R_B$ through R_2 and produces a high logic level at its corresponding output terminal

$$e_{o2} = e_{p2} + \frac{e_{p2} - V_-}{R_B} R_H$$

Thus, a voltage difference of $(R_H/R_B)(e_{p2} - V_-)$ separates the two logic levels. Note that V_- here will be a negative value, and its presence in these equations increases rather than decreases the voltages expressed. The operation of the complete Fig. 10.11 circuit parallels that described for the two-amplifier example here. The amplifier having the largest input voltage will control e_{fb} and produce a high logic

242 Chapter Ten

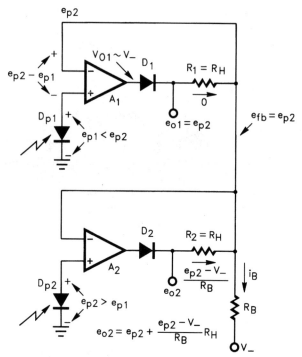

Figure 10.12 In a two-amplifier simplification of the Fig. 10.11 circuit, a larger diode signal e_{p2} overrides a smaller one e_{p1} to produce a high output state at e_{o2} and a low state at e_{o1}.

level at its corresponding e_o output terminal. All other amplifiers will produce low logic levels.

Diode matching and the amplifiers' offset and gain errors determine the accuracy of this digital position sensor. Together, they produce a net input error voltage $V_{i\epsilon}$. Diode responsivity mismatches make the light-to-voltage conversions differ for the diodes of an array, and for a given illumination condition, this produces a voltage mismatch error ΔV_p. Amplifier input offset voltage V_{OS} appears between the op amp inputs and adds directly to the diode error voltage. Similarly, amplifier gain error e_o/A_{OL} appears between an op amp's inputs, and the net input error voltage becomes

$$V_{i\epsilon} = \Delta V_p + V_{OS} + e_o/A_{OL}$$

where e_o is the output voltage of the op amp and not that of the maximum selector.

References

1. J. Graeme, "Use Op Amps to Design Optical Position-Sensing Circuitry," *EDN*, November 26, 1987, p. 229.
2. G. Tobey, J. Graeme, and L. Huelsman, *Operational Amplifiers; Design and Applications*, McGraw-Hill, New York, 1971.
3. S. Gage, *Optoelectronics/Fiber-Optics Applications Manual*, 2nd ed., McGraw-Hill, New York, 1977, p. 4.3.
4. F. Daghighian, "Optical Position Sensing with Duolateral Photoeffect Diodes," *Sensors*, November 1994, p. 31.
5. J. Graeme, "Generalized Op Amp Model Simplifies Analysis of Complex Feedback Schemes," *EDN*, April 15, 1993, p. 175.
6. J. Graeme, *Designing with Operational Amplifiers; Applications Alternatives*, McGraw-Hill, New York, 1977.

Glossary

avalanche photodiode A photodiode intended to be operated on the verge of avalanche breakdown so that impact ionization produces carrier multiplication and current gain.

bootstrap A circuit technique that removes signal voltage swing from a source by driving the source's common return with the signal appearing at the source's output terminal.

built-in voltage φ_B The potential barrier of a semiconductor junction as established by diffusion of carriers across the junction under zero bias.

common-mode input capacitance C_{icm} The effective capacitance between either input of a differential amplifier and common ground.

common-mode input resistance R_{icm} The effective resistance between either input of a differential amplifier and common ground.

common-mode rejection (CMR) The logarithmic form of common-mode rejection ratio (CMRR) as expressed by CMR = 20 log (CMRR).

common-mode rejection ratio (CMRR) The ratio of the differential gain of an amplifier to its common-mode gain.

common-mode voltage The average of the two voltages applied to differential amplifier inputs.

compensation resistor A resistor added in series with the noninverting input of an op amp circuit to develop an offset voltage component with the input current there that compensates for the offset component generated by the current of the noninverting input.

composite amplifier An op amp circuit enclosing two or more op amps within a common feedback loop.

current-to-voltage converter The op amp circuit configuration consisting of the op amp and a single, negative feedback resistor for converting an input current signal into an output voltage signal.

damping ratio ζ A figure-of-merit representation of the response damping characteristic of a second-order system.

dark current I_D The reverse saturation current of a photodiode or that current which flows through the diode under reverse bias and in the absence of illumination.

decoupling phase compensation A phase compensation technique that decouples an op amp from a capacitive load through an isolating resistor and a feedback bypass capacitor.

difference amplifier An op amp with a feedback configuration that results in an output voltage proportional to the difference of two input voltages.

differential input amplifier An amplifier having two inputs of opposite gain polarity with respect to the output.

differential input capacitance C_{id} The effective capacitance between the two inputs of a differential amplifier.

differential input resistance R_{id} The effective resistance between the two inputs of a differential amplifier.

duo-lateral photodiode A lateral photodiode having two current-dividing resistive surfaces for two-dimensional position detection.

electromagnetic noise coupling The parasitic signal coupling from a magnetic source through mutual inductances.

electrostatic noise coupling The parasitic signal coupling by an electric field through mutual capacitances.

feedback factor β The fraction of an output signal fed back to the input in a feedback system.

flux ϕ_e The radiant flux energy of a light source in watts.

gain error signal The differential signal voltage developed between the two inputs of an op amp and equal to the output signal voltage divided by the amplifier's open-loop gain.

***I*-to-*V* gain** Current-to-voltage gain.

impact ionization The generation of free carriers through carrier collisions with semiconductor atoms.

input bias current I_B The dc biasing current drawn by both inputs of an op amp.

input capacitance *See* Common-mode input capacitance and Differential input capacitance.

input noise current i_{ni} The noise current drawn by both inputs of an op amp.

input noise voltage e_{ni} The input-referred equivalent noise voltage of an op amp that would reproduce the noise at the output produced by internal noise sources.

input offset current I_{OS} The difference between the two input bias currents of an op amp.

input offset voltage V_{OS} The dc voltage impressed between the inputs of an op amp that produces zero output voltage.

input resistance *See* Common-mode input resistance and Differential input resistance.

instrumentation amplifier A differential input amplifier with internal feedback committed for voltage gain.

integrator An op amp with a feedback configuration that results in an output signal proportional to the time integral of the input signal.

intercept frequency f_i That frequency at which an op amp circuit's $1/\beta$ response intercepts the op amp's A_{OL} response, marking the highest frequency for which the circuit's available gain supplies the feedback gain demand.

intrinsic layer A semiconductor layer containing no P or N doped atoms.

junction capacitance An equivalent capacitance that models the charge storage of a semiconductor junction.

lateral photodiode A position-sensing photodiode with a resistive surface path that divides the diode's output current into multiple portions according to the position of a light beam on the diode's surface.

loop gain $A\beta$ The excess gain available to supply increasing feedback gain demand.

maximum selector A device which selects the largest of a set of input signals and connects the selected signal to the circuit output.

noise *See* Input current noise and Input voltage noise.

noise density The spectral noise contained in a 1-Hz bandwidth at a given frequency.

noise gain peaking A noise gain phenomena common to photodiode amplifiers in which the gain supplied to the op amp's input noise voltage rises and peaks at high frequencies.

open-loop gain The ratio of an op amp's output signal magnitude to the signal magnitude appearing between the amplifier's inputs.

phase compensation Frequency response tailoring for the stability of a feedback system through the addition of response poles and zeroes that reduce high-frequency phase shift.

phase margin ϕ_m The margin separating the phase shift around a feedback loop from 360° at the frequency where loop gain equals unity.

photoconductive The current-output operating mode of a photodiode.

photodiode A semiconductor junction that converts the photon energy of light into an electrical signal.

photovoltaic The voltage-output operating mode of a photodiode.

PIN photodiode A photodiode formed with an intrinsic layer between the normal P and N layers to expand the diode's depletion region and increase spectral bandwidth.

radio-frequency interference (RFI) The higher-frequency form of electromagnetic noise coupling.

rate of closure The difference in slopes of the $1/\beta$ and A_{OL} responses at their crossing as expressed in decibels.

responsivity r_ϕ The light-to-current gain of a photodiode as expressed by $r_\phi = i_p/\phi_e$, where ϕ_e is the radiant flux energy of the light in watts.

slewing rate S_r The maximum rate of change of output voltage.

tetra-lateral photodiode A lateral photodiode having two orthogonal sets of output electrodes on one current-dividing resistive surface for two-dimensional position detection.

unity-gain crossover f_c The frequency at which the open-loop gain of an op amp crosses unity.

virtual ground A groundlike characteristic of the feedback input of an op amp at which feedback absorbs injected current without developing a voltage.

voltage follower The short-circuit feedback connection of an op amp which results in an output signal that follows the signal at the amplifier's noninverting input.

Index

Assembly precautions, 129, 166, 197, 198, 212
Avalanche photodiode, 9, 245

Bandwidth, photodiode amplifier, 28
   advantage, 50
   with bias, 63
   wiith bootstrap, 71, 77
   improvement, 63, 140, 150, 157
   op amp limit, 33
Bandwidth-versus-noise:
   with current gain, 161
   with a voltage amplifier, 141, 144
   with voltage gain, 152
Bias, photodiode, 2, 5, 63
   with a current-to-voltage converter, 65
   filtering of, 67
   improvements to, 67, 69
   leakage produced by, 65
   noise produced by, 65
Bootstrap, photodiode, 71, 77, 245
   amplifier, 71, 151
   bandwidth of, 74, 83
   buffer for, 78, 80
   in combination with the current-to-voltage converter, 77
Bypass, power supply, 165
   detuning of, 186, 188
   the requirement for, 166
   selecting, primary, 173, 177
   selecting, secondary, 179, 182, 183

Characteristic curves, photodiode, 5
CMR (*see* Common-mode rejection)

Coaxial returns, 217
   with a ground plane, 218
Common-mode:
   input capacitance, 245
   input resistance, 245
Common-mode rejection, 245
   of bias errors, 69
   of electrostatic coupling, 199
   of magnetic coupling, 212
Common-mode rejection ratio, 245
Common-mode voltage, 245
Compensation resistor, offset, 24, 245
Composite amplifier, 112, 245
Contact resistance, photodiode, 14
Coupling, parasitic:
   electromagnetic, 210
   electrostatic, 198
   power supply noise, 166
Current amplifier, 157
Current gain:
   avalanche photodiode, 9
   photodiode amplifier, 157
Current-to-voltage converter, 23, 30, 77, 245
   bandwidth of, 33
   differential-input, 199, 207
   response of, 31

Damping factor, 53, 246
Dark current, 6, 246
Decoupling, phase compensation, 121, 123, 246
Decoupling, power supply, 191
   alternatives, 191
   selecting, 195

Detuning:
  L-C resonance, 36, 75, 175
  power supply bypass, 186, 188
Difference amplifier, 246
Differential-input:
  capacitance, 246
  resistance, 246
Differential-input amplifier, 246
  basic, 199
  filtering with, 224
  improved, 207
  position sensing with, 227
Digital position sensing, 239
  setting logic levels for, 241
  signal processing for, 239
Duo-lateral photodiode, 15, 246

Electromagnetic coupling, 210, 246
Electrostatic coupling, 198, 246

Feedback, 38
  analysis, 38
  demand, 39
  factor, 38, 246
  net factor, 75
Filtering:
  active, 120
  of bias source, 67
  with C_f bypass, 108
  with a composite amplifier, 112
  with decoupling, 121
  with a differential amplifier, 224
  passive, 121
  with a voltage amplifier, 142
Flux, 3, 246

Gain, added:
  current, 157
  with a current amplifier, 157
  optimum value of, 142, 145, 155, 161
  with a tee network, 141
  voltage, 150
  with a voltage amplifier, 140
Gain-bandwidth-product, op amp, 51
Gain error:
  in position sensing, 222, 231

Gain error (*Cont.*):
  signal, 33, 246
Gain peaking:
  noise, 90, 97, 107, 165
  signal, 50, 53

High-gain amplifiers, 129

I-to-V gain, 246
Impact ionization, 10, 246
Input, op amp:
  bias current, 24, 246
  noise current, 89, 246
  noise voltage, 89, 246
  offset current, 24, 247
  offset voltage, 27, 247
Instrumentation amplifier, 207, 247
Integrator, 247
Intercept frequency, 40, 247

Junction, semiconductor, 1
  avalanche, 9
  capacitance of, 5, 247
  depletion region of, 1
  photo response of, 3
  PIN, 7
  (*See also* Photodiode)

Lateral photodiode, 11, 247
L-C tank:
  amplifier equivalent, 36, 75
  power supply line, 176
Leakage current, 6, 25, 65
Linearity, 21
Loop gain, op amp, 40, 247

Magnetic coupling, 210
Magnetic field generation, 216
Magnetic loop, 212
  identifying, 212
  matching of, 215
Maximum selector, 239, 247
Model, photodiode:
  basic, 4
  duo-lateral, 16
  lateral, 13
  tetra-lateral, 18

Mutual capacitance, 198
Mutual inductance, 210

Noise, 87
  1/f, 95
  current, 90, 93
  density, 88, 247
  floor, 95
  power supply, 166
  region, 96
  resistor, 89, 93
  rms, 92
  rms integral, 93
  signal-to-noise ratio, 89
  total output, 100
Noise analysis, photodiode amplifier:
  with C_f bypass, 110
  with a composite amplifier, 118
  with current gain, 163
  with a decoupling filter, 126
  with a differential amplifier, 202, 209
  with differential inputs, 202, 209
  simplified, 95
  summary, 99
  with a tee network, 138
  with a voltage amplifier, 146
  voltage component, 97
  with voltage gain, 155
Noise bandwidth:
  with C_f bypass, 110
  with a composite amplifier, 112
  with a decoupling filter, 121
  versus signal bandwidth, 112
Noise effects:
  combining, 99
  general, 87
  op amp, 94
  predominant, 102
Noise, external, 197
  electrostatic coupling, 198
  magnetic coupling, 210
Noise gain, 67, 88
  with C_f bypass, 108
  with a composite amplifier, 115
  with a decoupling filter, 122
  peaking, 90, 97, 247

Noise gain (*Cont.*):
  with a tee network, 130, 134
Noise, power-supply coupling, 166
  frequency response of, 168
  and frequency stability, 170
  mechanism, 167
  reduction, 165
Noise reduction, 107
  with C_f bypass, 108
  with a composite amplifier, 112
  with a decoupling filter, 121
  with a differential amplifier, 199, 207, 224
  with electrostatic shielding, 198
  of magnetic field generation, 216
  with magnetic loop reduction, 212
  with magnetic shielding, 210
  in the presence of multiple effects, 216
  trial-and-error approach to, 216
Noise-versus-bandwidth optimization:
  with a composite amplifier, 117
  with a decoupling filter, 123
Normalization, 231
  digital alternative to, 239
  with a divider, 232
  dynamic range of, 233
  with a multiplier, 234
  simplified, 237

Offset, 24
  in position sensing, 221
  reduction, 24, 25, 130
One-dimensional sensing, 222
Oscillation, 31, 41, 157, 172

Parasitic:
  capacitance, 32
  feedback loop, 171
  inductance, 166, 167
  resistance, 174, 179
Peaking (*see* Gain peaking)
Permeability, 211
Phase compensation, 247
  alternatives, 53, 55
  with a composite amplifier, 117

Phase compensation (*Cont.*):
    decoupling, 121, 124
    effect of, 42
    realizing small values of, 59
    requirement for, 36, 75, 84
    selection of, 49, 57, 77, 84
Phase margin, 247
Phase margin analysis, 47
Photoconductive, 7, 247
Photodiode, 1, 247
    avalanche, 9
    capacitance, 5, 64
    characteristic curves, 5
    contact resistance, 14
    dark current, 25
    duo-lateral, 15
    lateral, 11
    leakage current, 25, 65
    multiple-element, 10
    PIN, 7, 248
    step response of, 3
    tetra-lateral, 17
    time constant of, 3
Photodiode model:
    basic, 4
    duo-lateral, 16
    lateral, 13
    tetra-lateral, 18
Photovoltaic, 7, 248
PIN photodiode, 7, 248
Position sensing, 10, 221
    differential, 225, 227
    diodes, 10
    direct-displacement, 222
    gain error of, 221, 231
    normalized, 231
    offset error of, 221
    single-axis, 222
    two-axis, 229
Positioning distance, 13
Positioning resistance, 13
Power-supply bypass, 165
    detuning, 186, 188
    the requirement for, 166
    selecting, primary, 173, 177
    selecting, secondary, 179, 182, 183
Power-supply decoupling, 191

Power-supply decoupling (*Cont.*):
    alternatives, 191
    selection, 195
Power-supply noise coupling, 166
    frequency response of, 168
    and frequency stability, 170
    mechanism, 167
    reduction of, 165
Power-supply sensitivity, 168
Primary-supply bypass, 173, 177

Rate-of-closure, 41, 248
Resonance, 37, 174
    amplifier, 39
    bypass capacitor, 179
    detuning of, 37, 186, 188
    dual bypass capacitor, 186
    intuitive evaluation of, 176
    power-supply line, 173
Responsivity, photodiode, 3, 248
RF detector, parasitic, 215
RFI coupling, 210, 215, 248

Second-order response, 53
Secondary bypass, 179, 182, 183
Shielding:
    electrostatic, 198
    magnetic, 210
Slewing rate, 248
Stability, 42, 171

Tee network, 25, 59, 130
    design, 137
    noise gain, 130, 134
    optimization, 134, 135
    resistance noise of, 135
    signal gain, 130
    voltage noise gain, 134
Tetra-lateral photodiode, 17, 248
Transimpedance amplifier (*see*
        Current-to-voltage converter)
Two-dimensional sensing, 229

Virtual ground, 33, 248
Voltage gain, 140, 150

Wideband amplifiers, 63

ABOUT THE AUTHOR

Jerald Graeme is an internationally acknowledged authority on electronic amplifiers. He is currently principal engineer with Gain Technology Corporation, where his activities include the design of high-speed analog integrated circuits. He previously spent close to three decades with Burr-Brown Corporation, where he participated in the design of op amps, instrumentation amps, variable-gain amps, V/F converters, two-wire transmitters, and analog multipliers.

He has written three op amp books for McGraw-Hill and more than 100 articles for publications such as *EDN* and *Electronics Design*. He is the holder of eight U.S. patents, and in 1993 was named Innovator of the Year by *EDN*.